Don's Theory

on

Aerodynamic Lift

1st edition : June 2018
Authored and published in Sri Lanka by

Air Commodore KGDN Jayasinghe
MSc (Def & Strat), BSc (Def Mgt), psc

Tele 0094714163720
Email : nihalj68@gmail.com

In part fulfillment of the doctoral thesis (PhD)
Faculty of Graduate Studies
General Sir John Kotelawala Defence University
Colombo, Sri Lanka

ISBN 978-955-35757-0-8

Printed by
Dissananda Press
No 35, Nawala Road, Nugegoda
Sri Lanka

ACKNOWLEDGMENT

The atmosphere in the Sri Lanka Air Force (SLAF) became my oasis of research into aviation since the year 2001 in which I embarked on a project with a small team of engineers to design and build an ultra-light aircraft. Then Director of Aeronautical Engineering Air Vice Marshal (retired) ODNL Perera offered his fullest support to the project by authorizing the use of human resource, materials and facilities available at the Aeronautical Engineering Wing (AEW) located at the SLAF Base, Katunayake, Sri Lanka. AVM Perera even authorized the use of 27 bhp SACHS- 350 engine fitted to Israeli built 'Super Scout' UAV to power the ultra-light. Wing Commander (retired) Varuna Senarathne who is now a Captain in Sri Lankan airlines was my co-partner in designing the ultra-light. My highest gratitude is expressed to the 13602 Sergeant Wijesuriya J who worked at the AEW as a specialized structural technician at the time, who built the ultra-light entirely by himself. Group Captain Mohan Balasuriya, Group Captain Udula Wijesinghe and Mr Viraj Fernando were close associates who provided their unconditional support in executing the project in its entirety.

The number of discussions, amount of thinking and dreaming about the methodology of crafting the intended product enticed me to observe everything I see in birds and in aircraft that creates the magic of aerodynamic lift. Renowned model aircraft builder in Sri Lanka Mr Viraj Fernando suggested using a 'Clark Y' wing profile to our ultra-light which he believed as one of the cleanest aero foils that produces a significant lift. We did not have access to any reliable methodology to check it on our own other than believing what Mr Viraj proposed trusting on his credentials. The 'Clark Y' profile was indeed easy to manufacture owing to its cambered upper surface and the flat bottom. Finally, the project 'Centenary X' was completed and it was successfully flown first time publicly by Wing Commander (retired) Varuna Senarathne on 17 Deember 2003 at the airfield, Sri Lanka Air Force Base, Ratmalana, during the celebrations held to mark the 100 years since the first powered flight by the Wright brothers in 1903.

The second air vehicle building project was started in the year 2009, shortly after the end of war against the Liberation Tigers of Tamil Elam (LTTE). SLAF used Israeli built 'Super Scout' 'Searcher MK II' and 'Blue Horizon II' UAV systems extensively to combat the terrorism since 1996. Having trained on the engineering aspects of UAV systems by Israeli Aircraft Industries (IAI) many times and also having experience working with the UAVs for nearly 14 years, induced the confidence in me to believe that building a UAV system on our own was a possibility. Air Chief Marshal (retired) Roshan Gunathilake who was the Commander of the Air Force at the time offered his fullest support to start the project. Group Captain Mohan Balasuriya, Wing Commander (retired) Thejeka Wanigasekera and Engineer Salinda Thennekoon were my co-partners in building and flying the first UAV, UX 001. The UAV was completely designed and built indigenously. As much as, I was a staunch believer of the performance of 'Clark Y' wing profile, I did not have access to any practical methodology to prove it. In the year 2009, aviation professionals venerated the 'Longer Path' or 'Equal Transit Time' theory as the most accurate explanation of the 'aerodynamic lift', though we could not find any plausible set of equations that could be used to calculate the generated lift forces in different flight profiles. Subsequent to the building of the first UAV, our team researched vehemently on the subject while building and flying a couple of different UAV platforms. My continuous indulgence and dreaming about the ways of optimizing the air vehicle performance in terms of endurance often led me to think and review the lift theories in the contemporary aviation annals. I often talked with pilots and tried to visualize the way they feel the magic of flight. Yet, I continued to worry about the lack of straight forward explanation which is simpler to understand. In the beginning of 2017, my relentless interest in the phenomena of 'aerodynamic lift' led me to enroll in a PhD program at the Kotelawala Defence University with the expectation of finding something useful related to the subject. There, I was assigned with two brilliant supervisors Dr Nirosh Jayaweera and Dr Saliya Jayasekera from the Aeronautical Engineering Department at the University of Moratuwa. Both scholars on the subject, spent many hours assisting me to achieve my objective of finding something new related to my study. Dr Nirosh and Dr Saliya deserve my sincere gratitude for their guidance in formulating and publishing my work.

One needs to understand the theory of circular motion and the three Newtonian laws in depth in order to understand the Don's Theory on aerodynamic lift. Flight Lieutenant Shanake Sampath ,the brilliant materials engineer who is a proud product of University of Moratuwa joined the UAV Research Station in latter part of 2017. He worked with me ceaselessly addressing many aerodynamic issues of the Unmanned Aerial Vehicles that the UAV Research Station was building. Engineer Sampath and myself would have spent many days carrying out analysis on the subject of aerodynamic lift before the light was shed on the new theory that I publish in this book. I would like to express my sincere gratitude and give the due credit to Engineer Sampath who was a great source of support in completing my initial research on the subject. 33462 Corporal Kumara AAA who generated the graphics of the book clearly depicting the phenomena was also a great source of strength that enabled me to complete the work to the expected standard.

It is also my profound responsibility to acknowledge the support and guidance extended by the Director General Engineering of SLAF Air Vice Marshal Andrew Wijesuriya, Director Aeronautical Engineering of SLAF, Air Vice Marshal Dushyantha Ratnayake and Commander of the Sri Lanka Air Force Air Marshal Kapila Veediya Bandara Jayampathy in making this work a reality.

CONTENTS

LIST OF TABLES

LIST OF FIGURES

Chapter One

EXISTING THEORIES ON AERODYNAMIC LIFT

Introduction

The phenomenon of aircraft heavier than air flying in the skies has multiple explanations developed by physicists, scientists and mathematicians. Unfortunately, many of the theories found in encyclopedias, on web sites and even in some text books are incorrect, causing unnecessary confusion to students NASA, (2018). Above statement appearing in the website of the '*Glen Research Center*' at National Aeronautics and Space Administration (NASA), USA as at 31st May 2018 is directly repudiating the most widely accepted explanation for the lift generation phenomena in aircraft wings which is also known as '*Longer Path*' or '*Equal Transit Time*' theory. The '*Lift*' depends primarily on the air density, square of the velocity, air viscosity, compressibility, the surface area over which the air flows, the shape of the body and the inclination of the body to the airflow. Since the beginning of modern aviation in 1903, the proponents who argue on the '*Theory of Aerodynamic Lift*' are divided into two camps. Those who support '*Bernoulli*' position argue that lift is generated by a pressure difference across the wing. The '*Newtonian*' camp argues that lift is the reaction force on a body caused by the deflecting flow of air. But it is important to note that Bernoulli never attempted to explain the aerodynamic lift. The attempt of Sir Isaac Newton by the proposition '*Newton's Sine Square Law*' could not fully encompass the phenomenon of lift into the theory. Therefore, the names of these two scientists are just labels for two camps according to the '*Glenn Research Center*' of NASA.

The most complex problem faced by the inquisitive mind about the magic of flight is the absence of an accurate theory that simply interprets the generated lift as a function of velocity, air density, wing area and shapes. The knowledge domain widely known as the '*Computational Fluid Dynamics*' or the '*CFD*' which is capable of calculating the pressure distribution curves that are practically generating around a moving aerofoil using the complex *'Navier – Stokes and Euler Equations'* has quenched the thirst of the aviation design industry. But, even a superfast computer may take a couple of days to resolve a set of equations which has to compute and solve millions of peripheral equations before the delivery of results. Yet, that result will only talk about a particular test segment. Ironically, it has no capacity to explain the phenomenon exploiting the general laws of physics. Hence, the objective of this book is to present the most simplified, yet acceptable explanation of the *'Aerodynamic Lift'* which will have a revolutionary impact on future designs, design techniques and even the way aircraft are flown.

'Longer Path' or 'Equal Transit Time' Theory

The qualitative definition of *'Longer Path'* or *'Equal Transit Time'* theory is derived by the application of *'Bernoulli Theory'* to the airflow above and beneath an aircraft wing. But, the latter part of the book explains, that the application of *'Bernoulli Equation'* into relationship between the *'Velocity and Pressure'* of a dynamic air flow above and beneath an aircraft wing does not support the *'Longer Path'* or *'Equal Transit Time'* theory. However, both engineers and aviators who embraced the aviation science as early as 1903 designed, built, explored and defined the generated lift of aircraft wings in association with 'Bernoulli Equation'. Excessive use of flat bottomed wings in early aircraft designs are evidence of the faith on *'Equal Transit Time'* theory. During the inverted flight at a given speed, the lift coefficient (C_L) is much lower resulting in an increased stalling speed; because of the lower lift coefficient (C_L), the wing must be set at a higher angle of attack than for the same speed in normal flight (*RAF Manual of Flying*). This statement is revealing. Even though, the year was 1955, in which the *'Longer Path'* or *'Equal Transit Time'* theory was venerated as the most accurate definition of the total lift generation, the statement admits that inverted wing is still creating lift to sustain the flight and the increased Angle of Attack (*AoA*) could compensate for a lowered coefficient of lift (C_L). If the, *'Bernoulli Equation'* is applied to define the generation of lift in an inverted conventional wing, the wing should be generating a negative lift rapidly losing the aircraft altitude which does not happen in real life. The exposition of these phenomena requires a systematic study and application of the basic laws of physics to understand the aerodynamic lift primarily in association with AoA and different convex and concave surfaces induced in to aircraft wings. The researchers in aerodynamics know that there is no unitary method of investigation in this field, *Lazar Dragos (2003)*. However, it is observed that the *'Longer Path'* Theory or *'Equal Transit* Time' theory is being taught for the student populations of both pilots and engineers as the most accurate principle which defines the generation of total lift and induced drag by aircraft wings. Figure 1:1 indicates the theory depicted in a diagrammatic form.

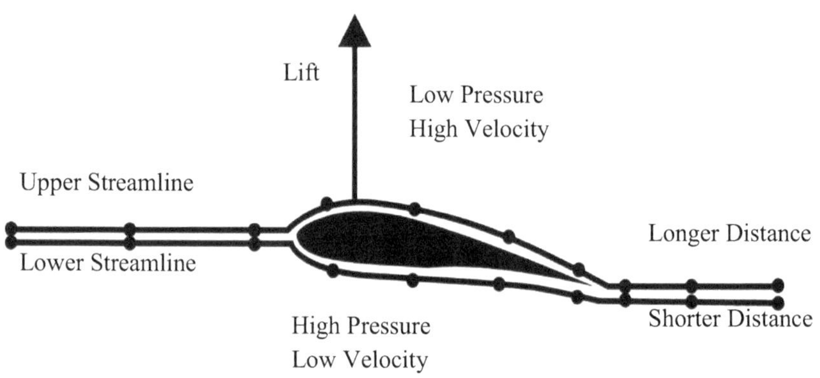

"Longer Path" or "Equal Transit Time" Theory

Figure 1:1- laminar boundary layers over and below the asymmetric aerofoil

(Source: NASA)

Air is actually considered a fluid in *'fluid dynamics'*. When modeling fluid movement, *Daniel Bernoulli (1738)* discovered that a faster moving fluid exerts less pressure than a slower moving fluid. This is the phenomena claimed to be acting on an airplane wing generating the lift and drag according to the *'Longer Path Theory'*. It is presumed that two air molecules travelling alongside at the tip of the leading edge must travel together as both molecules pass the trailing edge despite the path it takes to travel alongside the asymmetric aerofoil. The air molecules (fluid) moving over the top of the wing travels faster as it has to go around the longer curved surface in order to meet their travel buddies who took the shorter path below the wing surface. Hence, the speed of upper molecules increase which results in reduced dynamic pressure. The difference in dynamic pressure between the top and the bottom surfaces of the wing results in higher pressure acting at the bottom, thus pushing the wing upward which is termed as lift. But no scientific theory to explain why the separated air molecules at the leading edge must meet up again at the trailing edge which is the most significant flaw of the *'equal transit time theory'*.

Euler's Theory

Euler's Equations or the Euler's theory is another explanation which defines the aerodynamic lift around an airfoil. In a qualitative look at Euler's Equations, the movement of the fluid flow around the curved upper surface of the wing may be likened to that of a car going around a bend as depicted in the Figure 1: 2. As the car turns, the centrifugal force pushes the car away from the center of the turn but, the slant surface of the road is generating an equal and opposite force towards the center keeping the car in equilibrium which is known as the centripetal force. Similarly, as the fluid particle follows the cambered upper surface of the wing, the centripetal force acting on each little particle forces the particle to remain in the circular path. This force comes from a pressure gradient above the wing surface. Starting at the surface of the wing and moving up and away from the surface, the pressure increases with increasing distance until the pressure reaches the ambient pressure. Thus, a pressure gradient is created, where the higher pressures further along from the radius of curvature push inwards towards the center of curvature where the pressure is lower, thus providing the accelerating force on the fluid particle. Thus due to the curved, cambered surface of the wing, there exists a pressure gradient above the wing, where the pressure is lower right above the surface. Assuming a flat bottom, the pressure below the wing will be close to the ambient pressure, and will thus push upwards, creating the lift needed by the airplane.

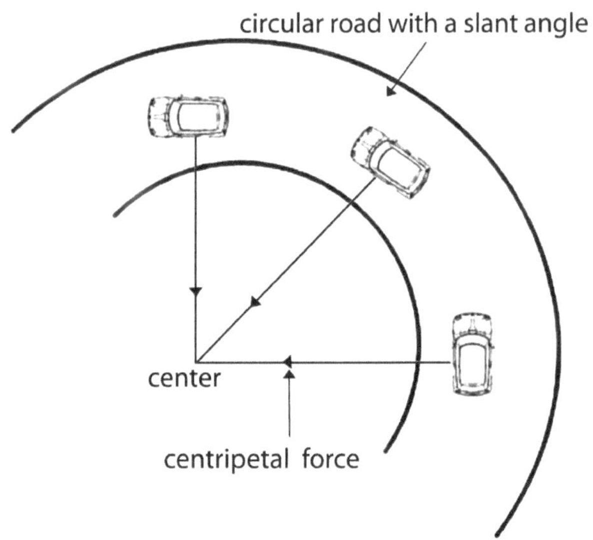

Figure 1:2 - centripetal force on a car accelerating around a curve

According to the Euler, it's the pressure gradient which is exerting the centripetal force on the air particles to move on the particular path. Basically, Euler's theory is also re-enforcing the belief that differential pressure generates the wing lift and it is therefore complimentary to the Bernoulli's Theorem. However, *'Euler's Equations'* are capable of interpreting complex conditions that are dependent upon many different variables. As a consequence, CFD applications are extensively using the Euler's equations to derive solutions.

Newton's Sine – Square Law[1]

The law generally known as the *'Newton's Sine–Square Law'* of air resistance refers to the force acting on an inclined flat plate exposed to the uniform air stream as depicted in Figure 1:3. It was much discussed in

[1] Theodore Von Karman, (2004), *Aerodynamics : Selected Topics in the light of their Historical Development,* Courier Corporation, P 09

connection with the problem of flight; in fact it cannot be found in Newton's works. It was deduced by other investigators based on a method of calculation which newton used for comparison of the air resistance of bodies of different geometrical shapes. In the thirty – forth proposition of his book he calculated the total force acting on the surface of spheres and cylindrical and conical bodies by computing and adding the forces caused by the impact of air particles, which supposedly move in a straight line until they hit the surface. The same idea applied to the calculation of the force acting on an inclined flat plate leads to the formula

$$F = \rho S v^2 \, Sin \, v^2 \alpha$$

Where 'ρ' is the density of the fluid, 'S' is the area of the plate; 'v' is the velocity of the plate and 'α' is the inclination angle. The force 'F' is directed normal to the plate. The quantity '$\rho S v \, Sine \, \alpha$' is evidently the mass flow in unit time through a cross section, '$S \, Sine \, \alpha$', equal to the projection of the plate perpendicular to the original flow direction. It is supposed that after the impact, the particles follow the direction of the plate. Then, the change of the momentum of the fluid mass hitting the plate in unit time has been calculated by multiplying the mass by the velocity component, '$V \, Sine \, \alpha$', created by the impact.

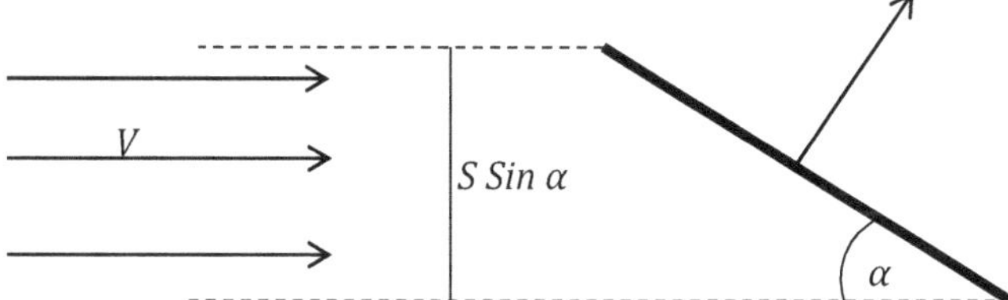

Figure 1-3, Newton's Sine-Square Law

It is pertinent to note that the dependence of the force on the angle of inclination was computed according to a particular assumption concerning the nature of the fluid flow, whereas its dependence on density, dimensions and velocity was determined by general principles of physics.

The Lift equation depicted in Figure 1:4 published by the NASA is widely accepted as the most accurate and authoritative equation to quantify the generated lift by an aircraft wing.

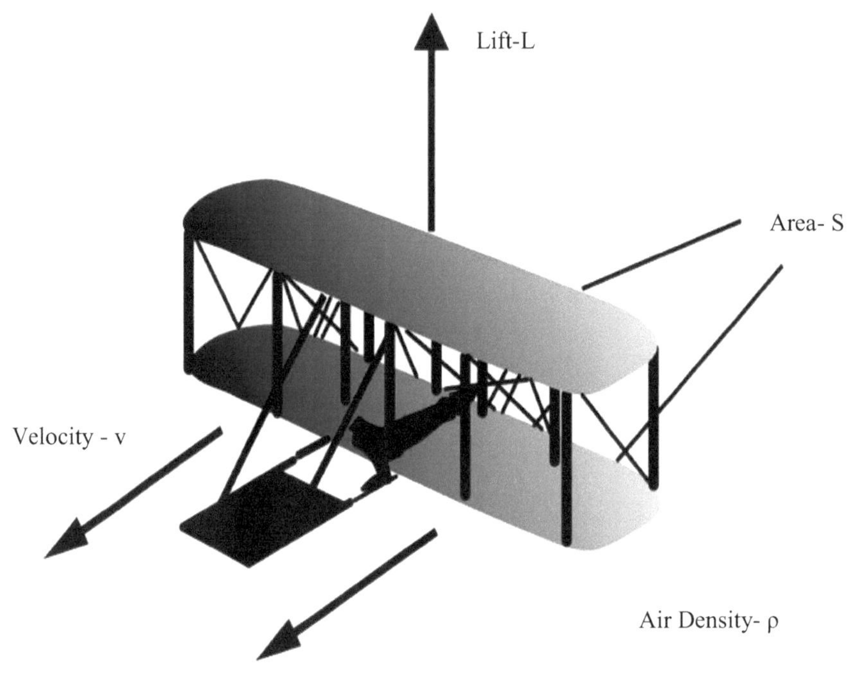

$$Lift = \frac{1}{2} \text{ Coefficient of Lift x Air Density x Velocity Squared x Wing Area } = \frac{1}{2} C_L \rho v^2 S$$

$$Lift(L) = \frac{1}{2} C_L \rho v^2 S$$

Figure 1: 4 – popular equation of aerodynamic lift

The origin of the Lift Equation

Obviously, the lift force (L) is the vertical vector value of all the reactionary forces generated by the velocity of the aircraft, air density, surface area and different geometrical shapes of the wing. The single momentous flaw of this popular lift equation is the omission of the impact of different shapes incorporated in aircraft wings and other lift control devices.

Therefore, assuming that velocity, air density and surface area of the wing are the only determinants of the generated *'lift'*,

$$L \quad \alpha \quad \rho \text{ (air density)}$$

$$L \quad \alpha \quad v \text{ (velocity of the aircraft)}$$

$$L \quad \alpha \quad S \text{ (surface area of the wing)}$$

$$L \quad \alpha \quad \rho^x v^y S^z$$

$$L \quad = \quad (k)\,(\rho^x)(v^y)(S^z)$$

Also,

$$F \text{ (force)} \quad = \quad kgms^{-2}$$

Hence,

$$kg \times m \times s^{-2} \quad = \quad k \times (kg \cdot m^{-3})^x \times (m^2)^y \times (ms^{-1})^2$$

$$= \quad k \times kg^x \times m^{-3x+2y+2} \times s^{-z}$$

By mass (kg),

$$x \quad = \quad 1$$

By time (s)

$$-z \quad = \quad -2$$

$$z \quad = \quad 2$$

By length (m)

$$-3x+2y+z \quad = \quad 1$$

$$-3+2y+2 \quad = \quad 1$$

$$2y \quad = \quad 2$$

$$y \quad = \quad 1$$

By the application of values of x, y and z,

$$L \quad = \quad k\,\rho^1 v^2 S^1$$

$$L \quad = \quad k\,\rho\, v^2 S$$

Hence, the units of measurements of density, velocity and surface area taken to quantify the '*Lift*' through '*dimensional analysis*' derives the final outcome in Newton Force ($kgmS^{-2}$). The value of 'k' cannot be solved mathematically and therefore it is said that the values of 'k' is found only through practical experimentation for a given aerofoil. Established values of 'k' through such experimental methodologies are expressed as $\frac{1}{2}$ x *Coefficient of Lift* (C_L).

$$K = \frac{1}{2} C_L$$

$$L = \frac{1}{2} C_L \, \rho v^2 S$$

Above equation that derived through '*dimensional analysis*' is not a derivative of '*Bernoulli's Theorem*' and it has following flaws.

- **No acceptable theory to support the relationship of each variable**

    Non availability of a straightforward theory to explain the relationship of each variable in the $L = \frac{1}{2} C_L \, \rho v^2 S$ equation is a substantial problem for the aircraft designer. The only variable that can be manipulated in a design according to the equation is the 'S' (area of the wing). Therefore, any manipulations done in shapes (convex or concave), thickness, incident angles etc. are without any straightforward explanations.

- **The values of 'C_L' could only be obtained through experimentation**

    Coefficient of Lift 'C_L' is a main manipulator of the equation, but its value can only be obtained through experimentation which is a significant flaw. Therefore, the exact variables that influence the value of 'C_L' and the behavioral pattern of those variables under different conditions are not known for many thousands of different aerofoils other than the CFD analysis data that have no theoretical explanations attached to them.

- **The AoA is not included as a variable in the equation**.

AoA is a significant variable that influence the generated lift. The inclination or AoA determines the mass *(m)* of the deflected flow and it directly influences the amount of transferred energy to the wing surface area. However, the lift equation has only the air density *(ρ)* as a variable which varies along with the altitude, but not with AoA at a given elevation. If the deflected air mass *(m)* could be included in the lift equation, it represents the deflected air volume, air density (function of temperature and humidity) and surface area over which it flows becoming the most significant variable. Therefore it is an absolute necessity that AoA or any other directly related parameter is represented in the Lift equation.

- **The equation doesn't specify the amount of lift generated from upper and lower surfaces separately**.

It is obvious that the lift equation $L = \frac{1}{2} C_L \rho v^2 S$ considers the surface area of the wing as a one unit that lifts up from the differential pressure acting from its bottom which is also a significant flaw. This perception has led to other considerations such as trailing edge vortices generated due to the clash of two pressures over the upper and lower wing surfaces and span wise airflow from root to the tip. Also, if an high pressure envelop is generated underside of the fast moving wing due to the differential velocities, dissipation of the high pressure to the atmosphere below the surface should be the first occurrence before it pushes up the wing generating an useful lift force upwards. There is no possibility to contain such generated pressure and project it only upwards towards the wing. Hovercraft is an example for this argument. The pressure bubble of the hovercraft is contained by a skirting around the pressurized area and by the earth or water surface underneath which cushions the craft above the ground. In the absence of such an arrangement to contain any pressure above the ambient will dissipate to the atmosphere immediately.

However, it is a widely believed phenomenon that the differential pressures generated due to the velocities of air flow that passes over and beneath the wing surfaces create the lift. The *'Dynamic Pressure'* variation being the most thought about variable, making some analytical calculations to understand the behavior of the dynamic pressure and its conversion into force about an asymmetric aerofoil is helpful to further understand the blemishes in the theory.

Relevant physical quantities and their units of measurements

Velocity	$= ms^{-1}$
Acceleration	$= ms^{-2}$
Pressure	$= kgm^{-1}s^{-2}$
Density	$= kgm^{-3}$
Force (Newton)	$= Pressure \times Area = (kg\ m^{-1}s^{-2} \times m^2) = kgm^2s^{-2}$
Kinetic Energy	$= \frac{1}{2}mv^2 = (\frac{1}{2}kg \times ms^{-1} \times ms^{-1}) = \frac{1}{2}kg\ m^2s^{-2}$
Dynamic Pressure (P_d)	$= Kinetic\ Energy\ per\ unit\ volume$

$$= \frac{1}{2}\ \frac{kgm^2s^{-2}}{m^3} = \frac{1}{2}kgm^{-1}s^{-2}$$

Also, (P_d)	$= \frac{1}{2}\rho v^2 = \frac{1}{2} \times kg/m^3\ (ms^{-1})^2 = \frac{1}{2}\ kgm^{-1}s^{-2}$
Potential Head	$= Potential\ Energy\ per\ unit\ volume$

$$= \frac{mgh}{V} = \frac{kg \times ms^{-2} \times m}{m^3} = kgm^{-1}s^{-2}$$

Bernoulli Equation

In fluid dynamics, *'Bernoulli's Principle'* states that an increase in the speed of a fluid occurs simultaneously with a decrease in pressure or a decrease in the fluid's potential energy. In another words, the sum total of the *'dynamic pressure'*, *'static pressure'* and *'potential energy'* is a constant.

$P_{dynamic} + P_{static} + \text{Potential Head } (mgh) \quad = \quad \text{Constant } (C)$

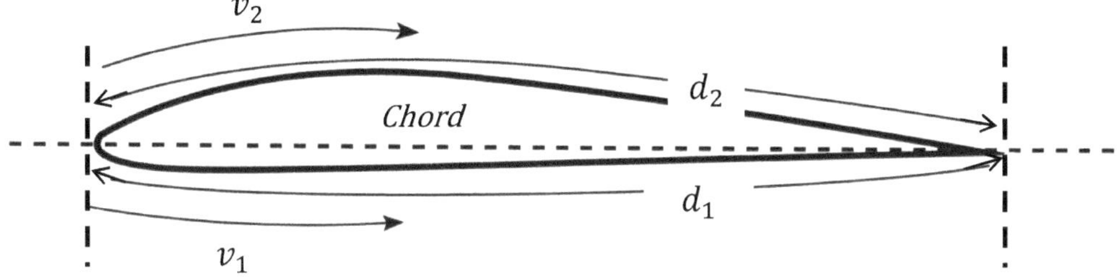

Figure 1:5 - asymmetric aerofoil

In the asymmetric aerofoil in Figure 1:5, v_1 and v_2 are the differential free airstream velocities while d_1 and d_2 depicts the respective distances that air stream travels over the top and bottom surfaces. P_1 and P_2 indicate the ambient static pressure.

$$d_1 < d_2$$

As time (t) taken to travel the distances are equal

$$\frac{v_2}{v_1} \quad = \quad \frac{d_2}{d_1} \quad\underline{\hspace{2cm}}\quad \text{①}$$

Application of the Bernoulli's Equation to the top and bottom surfaces of the wing

$$\frac{1}{2}\rho v_1{}^2 + mgh + P_1 \quad = \quad \frac{1}{2}\rho v_2{}^2 + mgh + P_2$$

Since the difference of potential head is negligible between the top and bottom surfaces.

$$\frac{1}{2}\rho v_1^2 + mgh + P_1 \; - \quad \frac{1}{2}\rho v_2^2 + mgh + P_2$$

$$\frac{1}{2}\rho v_1^2 + P_1 \quad = \quad \frac{1}{2}\rho v_2^2 + P_2$$

$$P_1 - P_2 \quad = \quad \frac{1}{2}\rho v_2^2 - \frac{1}{2}\rho v_1^2$$

$$\Delta P \quad = \quad \frac{1}{2}\rho (v_2^2 - v_1^2) \underline{\hspace{2cm}} \text{②}$$

Also,   Force (F) $=$ PA

Therefore, Lift $(L) =$ $\Delta\,P\,A$

By substitution of (2)

$$= \quad \frac{1}{2}\,\rho\,(v_2^2 - v_1^2)\,A$$

$$= \quad \frac{1}{2}\,\rho A\,\left(\frac{v_2 + v_1}{v_1}\right)\left(\frac{v_2 - v_1}{v_1}\right)v_1^2$$

$$= \quad \frac{1}{2}\,\rho A\,\left(\frac{v_2}{v_1} + 1\right)\left(\frac{v_2}{v_1} - 1\right)v_1^2$$

By substitution of (1)

$$= \quad \frac{1}{2}\,\rho A\,\left(\frac{d_2}{d_1} + 1\right)\left(\frac{d_2}{d_1} - 1\right)v_1^2$$

$$\Delta\,P\,A = \quad \frac{1}{2}\,\rho A\,v_1^2\left[\left(\frac{d_2}{d_1}\right)^2 - 1\right] \qquad\qquad (3)$$

The above derivative of the 'Bernoulli Equation' (equation 3) which includes the distances over the upper and lower surfaces of the wing as variables, mathematically prove two important paradoxes under following conditions.

a.   **Condition I:**   When $d_1 = d_2$ in a symmetrical wing, that wing generates zero lift. Therefore, an aircraft with a symmetrical wing cannot fly, according to the Bernoulli's equation.

b.   **Condition II:**   When an aircraft flies invert, its wings has to produce negative lift which does not happen in reality.

Also, it proves that the popular lift equation ($L = \frac{1}{2}C_L\,\rho v^2 S$) venerated by the engineers and pilots in the current context, is not a derivative of the Bernoulli equation. Yet, it remains as the most authoritative 'Lift Equation' backed by the NASA. All published literature including the publications by NASA state that C_L of the popular lift equation ($L = \frac{1}{2}C_L\,\rho v^2 S$) is derived experimentally for each different airfoil. Some scholars on the subject of aerodynamics claim that it is extremely difficult to obtain the values of ' C_L '. Also, C_L value is a variable combination of the Angle of Attack, Reynolds Number, temperature, Wing Chord etc. When considered the thousands of different wing profiles that are available today, it is unlikely that every manufacturer obtain the C_L/AoA curve through experimentation.

Dependence on powerful computers for CFD analysis to understand and interpret the behaviours of different profiles is the most reliable common practice adopted. But, the absence of a theory to explain the reasons for such behavior is the knowledge gap that requires filling up which will lead to new thinking in the domain of aircraft designs.

In the meantime, NASA has published a similar graph to the Figure 1:5, *'Lift/AoA curve'* which it recommends to refer as an approximation to the median values of many different aerofoils. But no authoritative source has published any algorithm to explain the relationships of different variables such as AoA, Reynolds Number and Wing Chord that are claimed to be the variables which control the values of C_L.

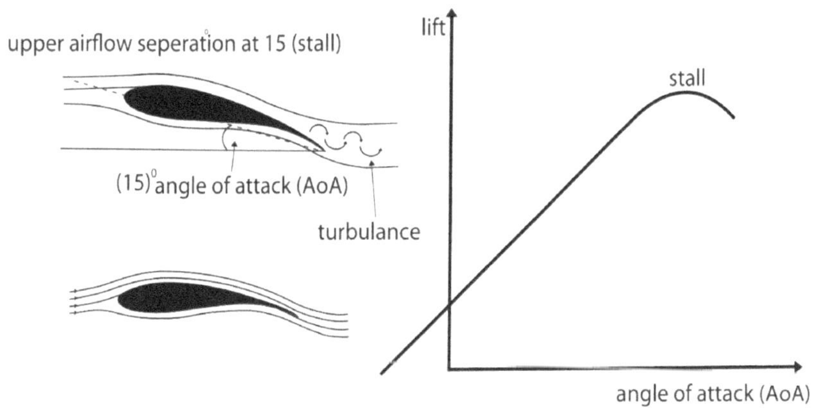

Figure 1:6 – correlation between Lift and AoA (Source: NASA)

Table 1:1 indicates the actual values measured from a series of different aircraft manufactured since 1935 and the ratio d_2/d_1 which is < 1 could be observed in all aircraft which gives a positive value to the differential pressure generated due to the Bernoulli Effect theoretically.

S/No	Aircraft	Year of Manufacture	Upper/Lower Lengths	$\dfrac{d_2}{d_1}$
1	Douglas DC-3 C-47 Dakota	1935	Upper length - 448 cm Lower length - 437 cm	1.025
2	Tiger Moth	1937	*Lower Wing* Upper length - 136 cm Lower length - 134 cm	1.014
			Upper Wing Upper length - 136 cm Lower length - 133 cm	1.022
3	DH 104 Dove	1945	Upper length - 253 cm Lower length - 247 cm	1.024
4	Austere J 1	1935	Upper length - 165 cm Lower length - 160 cm	1.031
5	Chipmunk	1946	Upper length - 188 cm Lower length - 179 cm	1.050
6	Balliol MK 2	1948	Upper length - 223 cm Lower length - 219 cm	1.018
7	Riley Heron	1950	Upper length - 311 cm Lower length - 308 cm	1.009
8	Pioneer MK 1	1950	Upper length - 261 cm Lower length - 253 cm	1.031
9	Mig 17	1950	Upper length - 268 cm Lower length - 262 cm	1.022
10	Beach craft E 18 (S – 9700)	1953	Upper length - 266 cm Lower length - 263 cm	1.011

11	Jet Provost T - 51	1954	Upper length - 211 cm Lower length - 206 cm	1.024
12	HS 748- Avro	1958	Upper length - 821 cm Lower length - 819 cm	1.002
13	PT - 6	1958	Upper length - 200 cm Lower length - 198 cm	1.010
14	Sky master	1961	Upper length - 183 cm Lower length - 181 cm	1.011
15	Siaimarchetti SF 260	1964	Upper length - 139 cm Lower length - 137 cm	1.014
16	Pukara	1966	Upper length - 230 cm Lower length - 226 cm	1.017
17	FT - 5	1966	Upper length - 208 cm Lower length – 205 cm	1.014
18	Centenary X	2003	Upper length - 133 cm Lower length - 131 cm	1.015

Table 1:1- measured chord wise distance ratios of upper *(d₂)* and lower *(d₁)* surfaces in 18 different aircraft

The mode of the above data set *(1.014)* could be taken to perform the approximate calculations of the differential pressure which acts upwards from the bottom of an airfoil section that is moving through the air at a considerable speed according to the theory. Also above measured data set indicates that the d_2/d_1 ratio is significantly smaller in magnitude.

Therefore, the value of $\Delta P A = \frac{1}{2} \rho A v_1^2 \left[\left(\frac{d_2}{d_1} \right)^2 - 1 \right]$ is always too insignificant to mathematically prove the adequacy of the differential pressure $(\Delta P A)$ generated underneath the wing due to differential velocities could deliver the required lift force in accordance to the *'Bernoulli Theorem'*.

Chapter Two

CIRCULAR MOTION

Newtonian Laws

Sir Isaac Newton first presented his three laws of motion in the *'Principia Mathematica Philosophiae Naturalis'* in *1686*. Since then, the Newtonian laws contributed to take giant leaps in physics by developing meaningful relationships among mass, length and time. Specifically, the ability of Newtonian laws to explain the complex relationships associated with velocity, force and acceleration are simple, yet profound. Newtonian laws have given a new meaning to the vector and scalar quantities in mathematically interpreting them.

First Law of Newton

Every object will remain at rest or in uniform motion in a straight line unless compelled to change its state by the action of an external force

Second Law of Newton

Force equals mass times acceleration (F=ma)

Third Law of Newton

Every action has an equal and opposite reaction

Velocity

Velocity is defined as the vector measurement of the rate of change and the direction of the motion ($\Delta d/\Delta t$). The scalar magnitude (absolute value) of the measurement is speed. The SI (international) units of velocity are meters per second (ms^{-1}).

Force

Force is a quantitative description of an interaction that causes a change in objects motion. An object may speed up, slow down or change direction in response to a force. Objects are pushed or pulled by the forces acting on them. SI unit of measurement of force is Newton and it is the force required to give a mass of *01 kg* an acceleration of *01 m* per second per second ($kgms^{-2}$).

Acceleration

Acceleration is the rate of change of velocity, either in its magnitude or in its direction, or both. Even if an object is moving at a constant velocity, but changes its direction only, that object is accelerating. The SI (international) units of acceleration are ms^{-2}.

Acceleration in uniform circular motion

No object or mass is naturally inclined to travel on a circular path unless acted upon by an external force in accordance with *first law of Newton.* Therefore, any object that travels on a circular path has to be continually acted upon by an external force in order for that object to remain in a circular orbit changing the direction of its velocity continually. However, its velocity is tangential to the center seeking force (centripetal) which is changing the direction of the velocity of the object continuously sustaining the circular motion. In uniform circular motion, the direction of the velocity changes continuously, which generates acceleration towards the center of the circle, even though the magnitude of the velocity is constant as indicated in Figure 2:1.

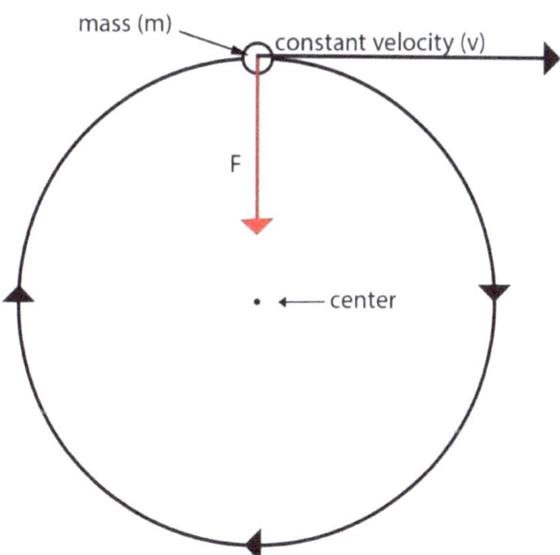

Figure 2:1 – center seeking force (centripetal) in circular motion

Centripetal acceleration of circular motion

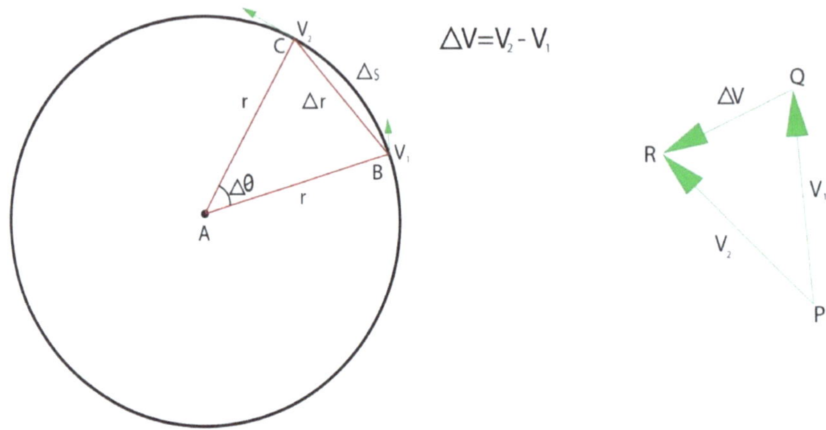

Figure 2:2 - vector diagram of circular motion

The directions of the velocity of a mass at two different points on the circle are considered, the change in velocity, $\overrightarrow{\Delta v}$ is seen to point directly toward the center of the circle according to the velocity vector diagram PQR depicted in Figure 2:2. Angle is measured in radians and since the radian $\Delta\theta$ is a small angle, Δs and Δr are also small values that could be considered equal.

Therefore, $\Delta s = \Delta r$ and *Distance = Time x Velocity*

$\Delta s = v\,\Delta t = \Delta r$

$\Delta\theta = \dfrac{\Delta r}{r}$ Δr, considering the *PQR* velocity vector diagram $\Delta\theta = \dfrac{\Delta v}{v} = \dfrac{v\,\Delta t}{r}$

$\dfrac{\Delta v}{\Delta t} = \dfrac{v^2}{r}$ = Acceleration *(a)* or more accurately centripetal acceleration *(a_c)*

Hence the Centripetal Force *(F_c)* = $ma_c = \dfrac{mv^2}{r}$

$$F_c = \frac{mv^2}{r}$$

Centripetal Force

Centripetal means *"toward the center"* or *"center seeking"* force. The centripetal force is the center seeking force exerted on a mass engaged in a circular motion. A force cannot be generated if there is no acceleration. A mass accelerating when it is in circular motion at constant velocity is counter-intuitive at first because we forget that changes in the direction of a moving mass, even if the mass is maintaining a constant velocity is counted as acceleration. Imagine swinging a ball attached to a rope in a circular motion. The ball is in a constant acceleration as the velocity of the ball is changing its direction continually. The acceleration of the ball is caused by the net force acting on the ball. This net force is called the *'Centripetal Force'* and is always pulling or pushing the mass towards the center of the circle. It is obeying the second law of newton which says that the 'force exerted on a moving object is equal to the mass times acceleration *(F = ma).* If not for the centripetal force, the ball will move on a straight line. In other words, centripetal force is the tension on the rope. This centripetal force *(f) = m v^2/ r.* The increase of mass or velocity or decrease in radius will result in an increase of the centripetal force. Increase of the radius or decrease in mass or velocity will result in the decrease of centripetal force. The changes in the velocity are making the widest variations in the centripetal force as the value of *'f'* is increased exponentially with the increase of *'v'.* If the rope is cut, the ball will no longer travel in a circular path but will travel on a straight line perpendicular to the radius of the circle at the point of exit, obeying the Newton's First Law of inertia.

Centrifugal Force

According to the *'Newton's Third Law',* every action has an equal and opposite reaction. Then, the centripetal force should also have an equal and opposite reactive force fleeing from the center of the circle. Any object that enters into a circular motion is acted upon by an external force which keeps that object in its circular orbit by exerting a force towards the center. The *'apparent'* reactive force acts opposite to the centripetal force is also known as the *'Centrifugal Force'.*

Chapter Three

AEROFOIL IN MOTION THROUGH THE AIR

Atmospheric air

The atmosphere of earth is composed of a layer of gases, commonly known as air and is retained by earth's gravity. By volume, dry air contains *78.09%* nitrogen, *20.95%* oxygen, *0.93%* argon, *0.04%* carbon dioxide, and small amounts of other gases. Air also contains a variable amount of water vapor, on average around *1%* at sea level, and *0.4%* over the entire atmosphere. The atmosphere has a mass of about *5.15×10^{18} kg*, three quarters of which is within about 11 km *(6.8 mi; 36,000 ft)* of the surface. The mass represents the molecular weight of different gases in the air. Having a mass is indicative of its properties that are obedient to the Newtonian laws. The atmospheric air density becomes thinner and thinner with increasing altitude, with no definite boundary between the atmosphere and outer space. The Karman line, at 100 km *(62 mi),* is often referred as the border between the atmosphere and outer space. Atmospheric effects become noticeable during atmospheric re-entry of spacecraft at an altitude of around 120 km *(75 mi).* Several layers can be distinguished in the atmosphere, based on characteristics such as temperature and composition.

Air density and mass

Air density is the mass of air per unit of volume it occupies, and it is expressed in kilograms per cubic meter (kgm^{-3}) when using the metric system. The density of air at sea level is about 1.2 kgm^{-3} (1.2 gl^{-1} or 0.0012 gcm^{-3}). Atmosphere consists of air molecules and its density decreases as the altitude increases. The lift of an aircraft wing, the aerodynamic drag of an aircraft, and the thrust of a propeller blade are all directly proportional to the air density. Similarly, the horsepower output of an internal combustion engine is also related to the air density. It is important to note that water vapor in the air causes a decrease in air density. Therefore, on a humid day, a wing has less lift.

Aerofoil

A streamlined surface designed in such a way that air flowing around it produces useful force *(NASA).* In another words, aerofoil is the term used to describe the cross sectional shape of an object that, when moved through the medium of air creates aerodynamic forces. These forces could be used in different magnitudes and combinations to produce the aerodynamic lift and control a flight in motion.

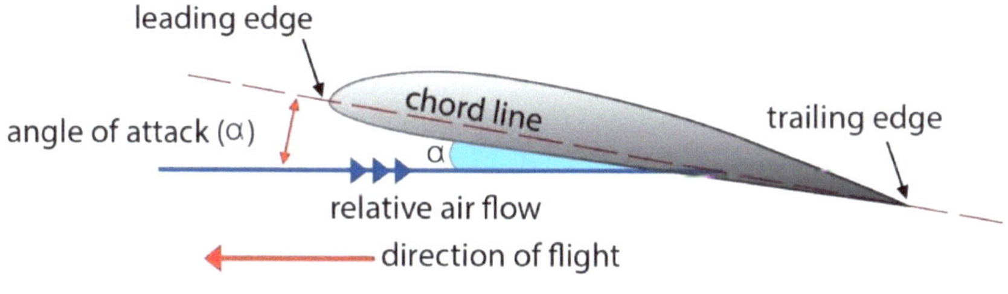

Figure 3:1- aerofoil

Leading edge and trailing edge

While the forward most point of the aerofoil is termed as the '*Leading Edge*', the aft most point is termed as the *'Trailing Edge'* as indicated in Figure 3:1.

Wing chord line

The straight line drawn from the leading to trailing edge of the aerofoil is called the '*Wing Chord Line*', denoted by the symbol *'C'*. The chord line cuts the aerofoil into an upper surface and a lower surface.

Angle of Attack (AoA)

As an aerofoil moves through the air, the wing is inclined to the flight direction at some angle. The angle between the chord line and the flight direction is called the angle of attack and has a direct effect on the lift generated by a wing.

Relative air flow

The relative air flow, also commonly called the relative wind is always acting against the airplane flight path and is considered to be relative because its motion is referred in relation to the direction of the flight path.

Change of aircraft pitch to maintain a constant angle of attack

Figure 3:2 illustrates a flight path in which the aircraft is in a decent, level and a climb. In these three situations the pitch of the aircraft is adjusted in a way to maintain a constant angle of Angle of Attack (*AoA*).

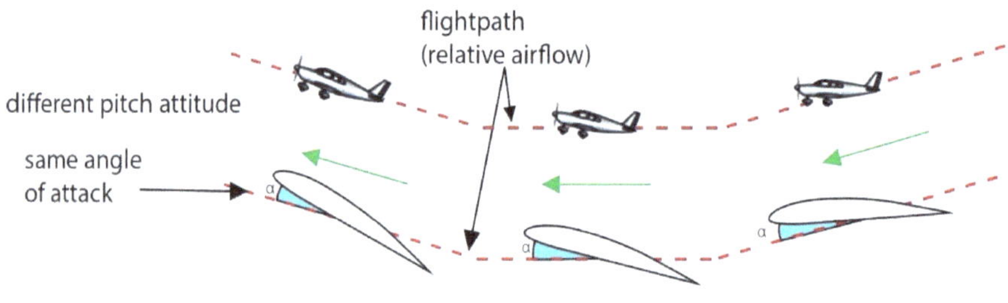

Figure 3:2 – constant angle of attack (AoA)

Change of angle of attack to maintain a constant pitch

The depiction on the Figure 3:3 indicate a situation in which the air craft pitch remains a constant, but its angle of attack changes as the aircraft descends, levels off and climbs. Therefore, relative airflow can be defined as the airstream parallel to the flight path or direction.

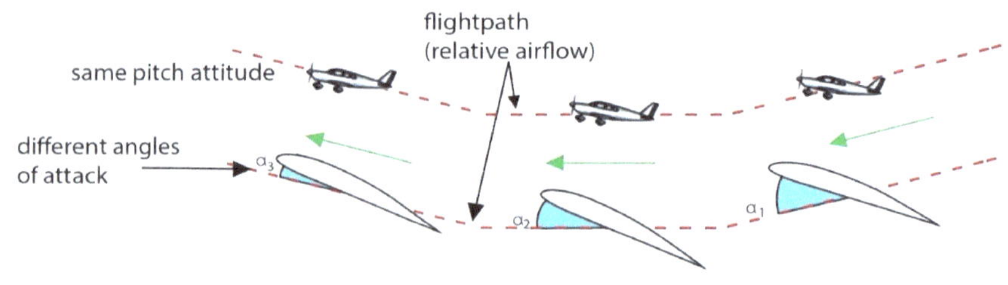

Figure 3:3 – Constant pitch

Coanda effect

The 'Coandă Effect' is the phenomena in which an air flow attaches itself to a nearby surface and remains attached even when the surface curves away from the initial air flow direction. However, the separation starts for angles of attack above 15º.

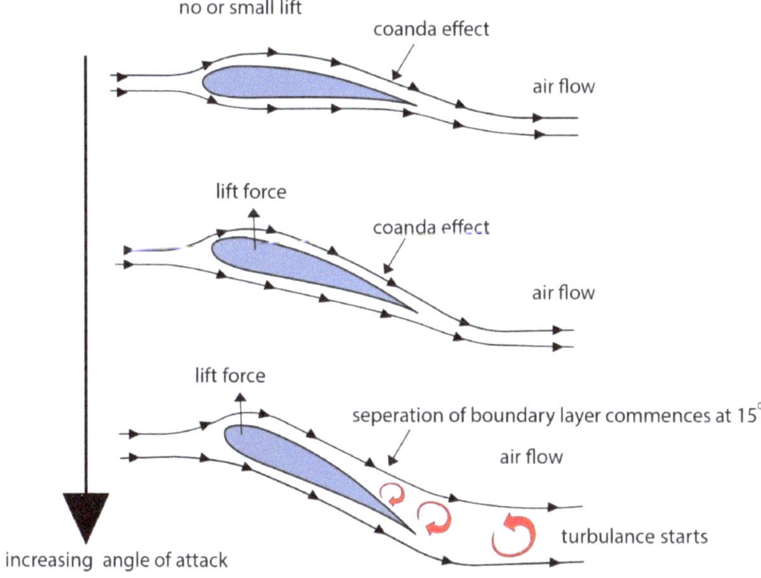

Figure 3:4 – coanda effect and progressive separation of boundary layer with increasing AoA

Air flow about an aerofoil

The air flow above an aerofoil is gushing in stream lines when the flow is not disturbed by turbulence. The camber of the indicated aerofoil in Figure 3:5 reaches its maximum height at the point 'D'. The air stream over the upper surface of the airfoil from 'A' to 'D' is kept turning (changing the velocity as the direction is continually changing and thereby accelerating) along the circumferential arc *(AD)* of the circle with radius 'r'. The stream lines are kept attached to the surface till point D since the surface AD is directly pushing the air stream on an upward direction. As the air molecules of the stream leaves the arc 'AD' and enters the arc 'DC' of the circle with radius 'r_2', it takes a downward turn. The airstream is kept attached to the 'DC' segment of the arc due to the *'Coanda Effect'* until it leaves the point *'C'*. The airstream (or the air molecules) are also kept accelerating from point 'D' to 'C' due to the continuous change of direction along the arc 'DC'. However, it has been proven experimentally that the flow over the upper surface after taking the downward turn, usually detaches from the surface for angles of attack above *15°* nullifying the *'Coanda Effect'* as depicted in *Figure 3:4*. In the terminology of aerodynamics, this phenomenon is termed as the *'stalling'* of the wing which is a significant loss of the generated lift.

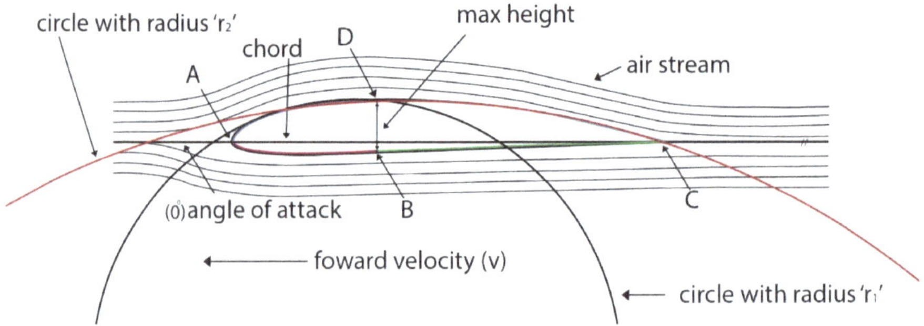

Figure 3:5 – circular motion of air molecules

Blemishes of existing aerodynamic lift quantification theories and equations

Having studied the existing philosophies, theories and concepts evolved since the time of *Da Vinci*, it is realized that a well formulated generic model to explain and quantify the useful lift generated by an aerofoil section is not published by any authoritative source to date. The aviation community is mostly using the known equation, $L = \frac{1}{2} C_L \rho v^2 S$ which is largely dependent upon the variable C_L. The values of C_L are dependent upon the following.

a.   Reynolds Number

b.   Ambient Temperature

c.   Humidity

d.   Compressibility

When considered the nature of above variables that influences the C_L in accordance with the accepted theories, it is not easy to determine their involvement or behavior to quantify them mathematically other than CFD analysis. Many books on the subject say that values of 'C_L' are obtained only through experimentation. Therefore, following blemishes are inherent in the theoretical concepts and mathematical equations used to quantify the *'lift'* in the modern aerodynamics.

a.   Mathematical application of the '*Bernoulli Principle*' proves that it does not accurately quantify the generated lift through the equation,

$$\text{Lift} \quad = \quad \frac{1}{2}\, \rho A\, v_1^2 \left[\left(\frac{d_2}{d_1} \right)^2 - 1 \right]$$

b. $L = \frac{1}{2} C_L\, \rho v^2 S$ is not a derivative of '*Bernoulli Principle*'

c. $L = \frac{1}{2} C_L\, \rho v^2 S$

    1)   Cannot differentiate among different airfoils

    2)   The curvatures of the aerofoils are not taken into calculation

    3)   The changing mass of the deflected air volume through the changing of AoA is not taken into the account

    4)   'S' does not represent both sides of the aero foil.

    5)   Values of 'C_L' could be obtained only through experimentation

d.   '*Equal Transit Time*' or '*Longer Path*' theory is wrong.

Chapter Four

DON'S THEORY ON AERODYNAMIC LIFT

Circular motion of air molecules about an aerofoil

A streamlined aerofoil moving through the medium of undisturbed air is continually pushing and pulling the air molecules into circular motion in the form of air streams from their original state of equilibrium. In another perspective, it can be considered, that the aerofoil is stationary, but the air streams are fleeing past the aerofoil with a velocity equivalent to the Indicated Air Speed (IAS) of the air vehicle. Air streams made up of molecules that have a certain mass and are being deflected into circular motion from the original state of equilibrium forcefully by the convex and concave surfaces of the aerofoil. Aerofoils are designed in such a way, when they are moving through the medium of air, the surface contours either 'push' or 'pull' the air streams into circular motion. Airstreams transfer from one circular contour to another in the form of a boundary layer as depicted in Figure 4:1.

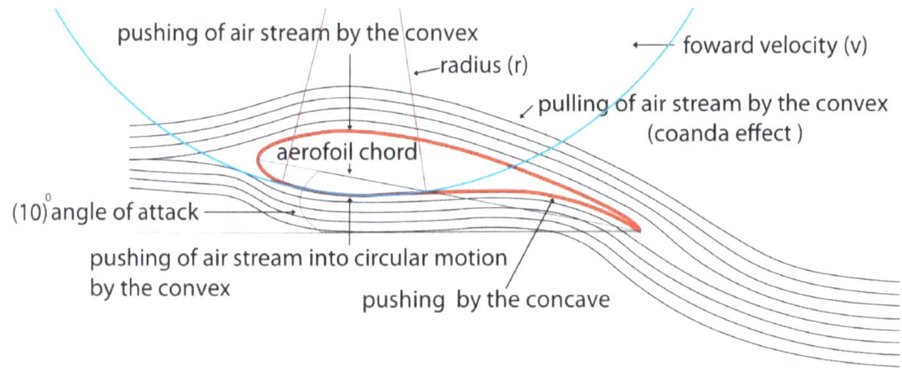

Figure 4:1- pushing and pulling of the air stream into circular motion

Entry and exit of air molecules into circular motion with different radii

According to the theory, as soon as any mass exits the circular motion, it travels on a straight line unless acted upon by an external force (in accordance to the first law of Newton). Also, a mass that exits the circular motion by breaking away from the centripetal forces of a particular circle and travels on a straight line can transfer into a circular motion with a different radius, if acted upon by centripetal forces of that different circle. Also, a change of momentum takes place in air molecules during the *'exit'* and *'entry'* from one circular motion to another due to the different radii of the two circles as depicted in Figure 4:2.

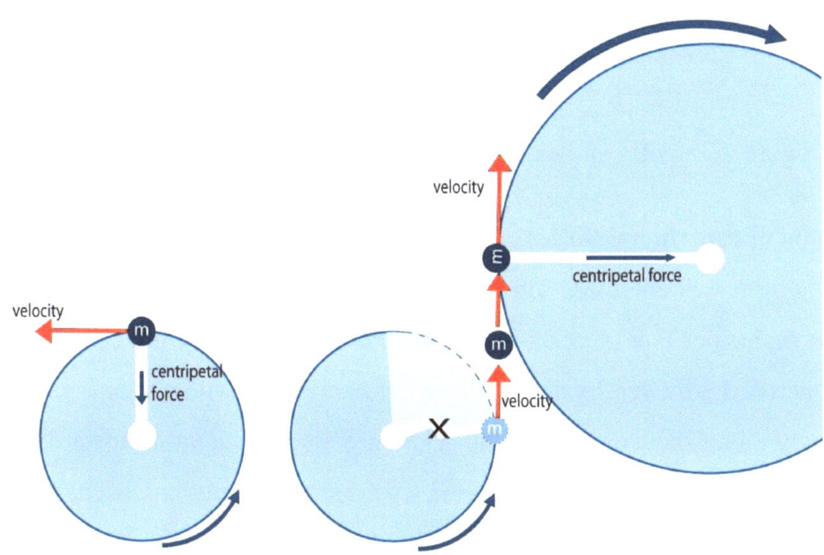

Figure 4:2 – exit and entry of a mass into circular motion

In most aerofoils, when the air molecules exit the circular motion around a particular convex or concave surface, it enters another with a different radius and an orientation. Therefore, air molecules that are speeding over the upper and lower surfaces of a wing profile in the form of air streams rotate over a minimum of two different circular surface segments before they escape at the trailing edge. As depicted in Figure 4:3, the molecule *'A'* enters the circular motion with radius (r_1) while the molecule *'B'* exits it and enters the into circular motion with radius (r_2). The molecule *'C'* completes the rotation around both convex and the concave and escapes at the trailing edge with a momentum that dissipates after a while.

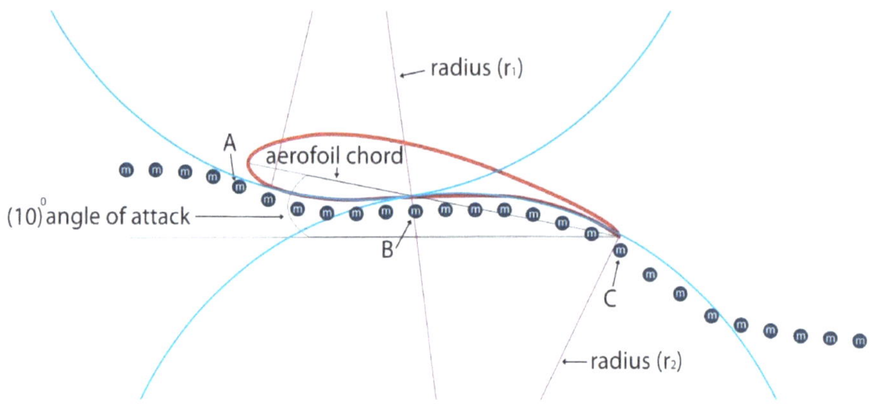

Figure 4:3 - entry and exit of air molecules into circular motion

Generation of centripetal forces

No object will enter into circular motion if not acted upon by an external force. Air molecules that are kept in circular motion forcibly by the convex and concave aerofoil surfaces by their respective curvatures without letting them travel freely along their original paths. They are also accelerating towards the center of the convex or concave along the way due to the continuous change of direction despite the constant velocity. These accelerating molecules that are having a certain mass are generating centripetal forces towards the center of the particular convex or concave they rotate around obeying the second law of Newton. This center seeking force is felt by the surface skin of the wing as an aerodynamic force (equal and opposite reactions to the corresponding centripetal forces) as depicted in Figure 4:4. The radius of the convex or concave surface that air molecules rotate around is a critical parameter which has the highest influence on the amount of centripetal force generated through the rotation as the generated force $(F) = \dfrac{mv^2}{r}$. The lift control devices of aircraft are controlling the radii of airstream rotation towards the trailing edge augmenting the lift in many folds which will be explained in a latter part of the chapter. The radius of a curvature integrated to an aerodynamic profile was never a dimension considered for any lift computation in the past.

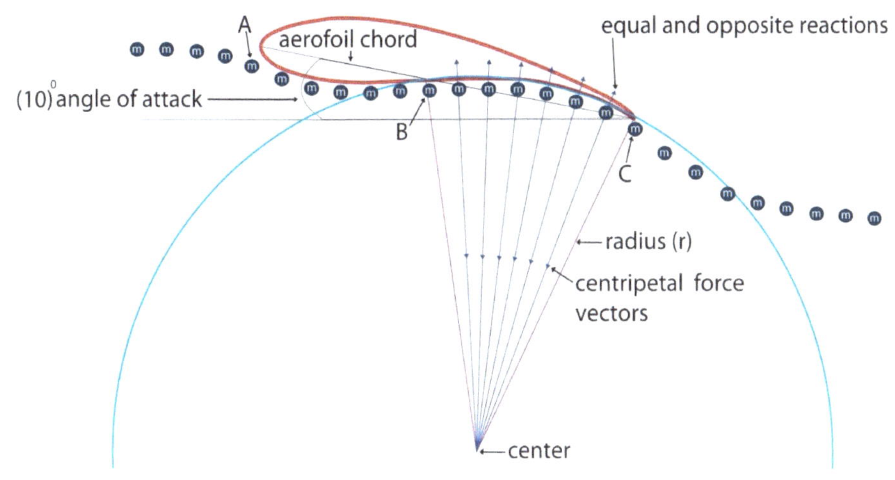

Figure 4:4 – acceleration of air molecules towards the center

'Push' and 'Pull' forces acting on the air stream

When the convex or concave surfaces are directly exposed to the upcoming air stream, those surfaces are forcibly *'pushing'* the air molecules into circular motion. On the contrary, when convex or concave surfaces are not directly exposed to the upcoming air stream, the *'coanda effect'* is *'pulling'* the air molecules into a circular motion as depicted in Figures 4:5

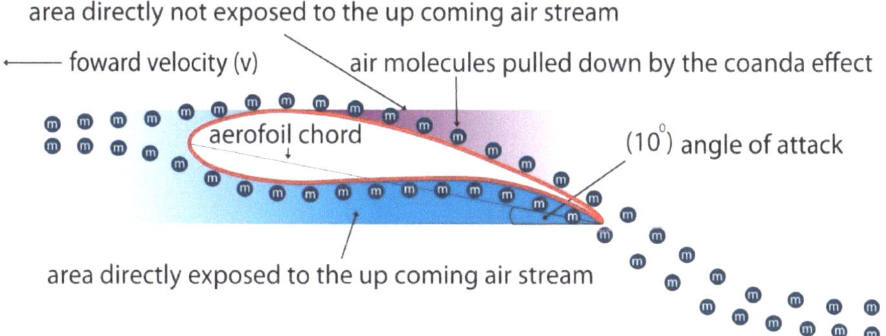

Figures 4:5 – direct exposure to the upcoming air stream

Coanda Effect explained

In accordance with the laws governing the circular motion, air molecules that leave the point 'D' in Figure 4:6, should be travelling on a straight line unless acted upon by an external force according to the first law of Newton. If it travels on a straight line, a vacuum has to be created between the airstream and the upper surface of the aerofoil after the point 'D'. Since such a vacuum cannot exist in equilibrium, it sucks the aerofoil surface and the airstream into it with an equal force.

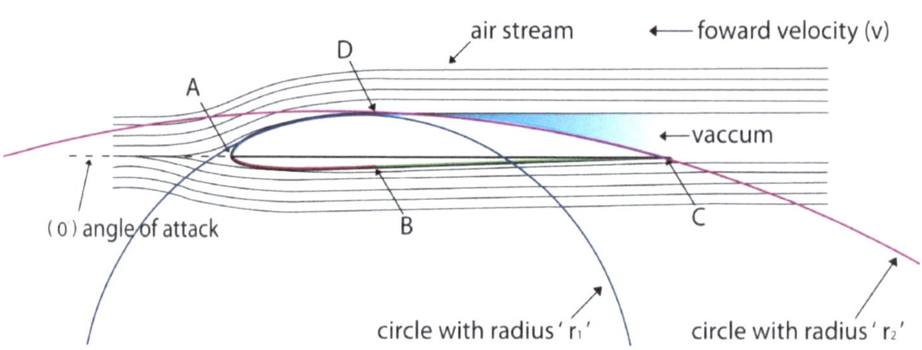

Figure 4:6 - theoretical depiction of the vacuum which does not occur in reality

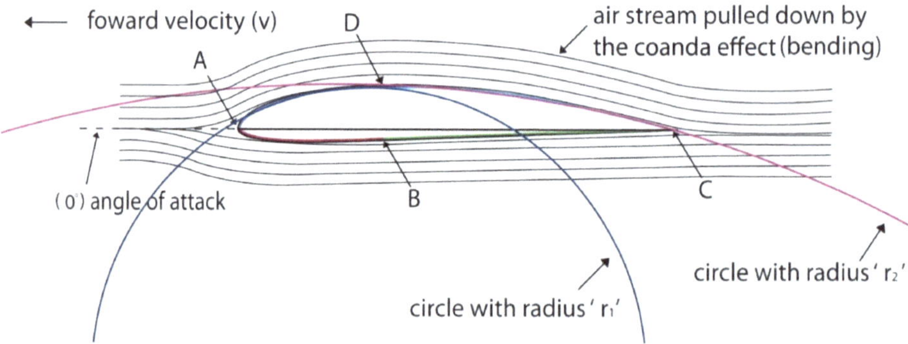

Figure 4:7 – bending of the air stream due to the suction of the vacuum

But, in reality the vacuum is never created. Before the vacuum is created, at angles of attack below 15^0- 18° , the airstream bends until it flows as a boundary layer along the convex 'DC' arc with a radius 'r_2' in accordance with the first law of Newton as depicted in **Figure 4:7**. The separation of the boundary layer after 15^0- 18° will be discussed in latter part of the chapter.

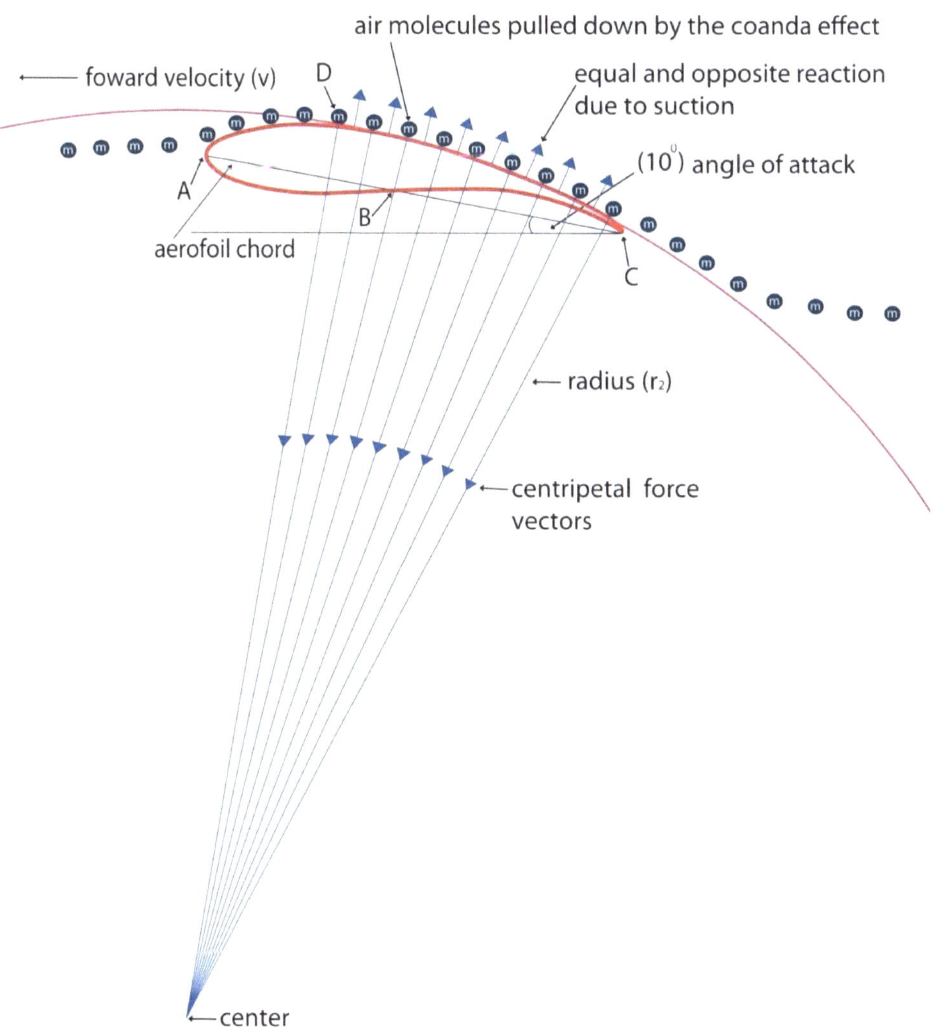

Figure 4:8 – lift by suction due to the 'coanda effect'

Lift generated by 'Coanda Effect'

As the air molecules continuously taking a turn along the convex *'DC'* due to suction effect of the potential vacuum, they accelerate generating centripetal forces towards the center of the convex with radius *'r_2'*. The equal and opposite reaction (suction) felt on the arc *'DC'* is equal to the sum total of centripetal forces exerted on the air mass (air stream) that curves along the surface *'DC'*. Therefore, the generated force due to the *'Coanda Effect'* is felt as a suction on the aerofoil instead of a pushing force as depicted in Figure 4:8.

Factors influencing the generated lift

It is clearly evident that the amount of lift generated from an aircraft wing is directly proportional to the following.

- Velocity of the aircraft
- Radii of the convex or concave surfaces incorporated to the wing
- Orientation of the convex or concave surfaces
- Mass of the rotated air volume
- Angel of Attack (AoA)

Velocity of the aircraft

Quantification of the lift can be done by direct application of the centripetal force equation $\left(F = \dfrac{mv^2}{r}\right)$ of a rotating mass. In this quadratic equation, the generated force increases exponentially with the increase of velocity.

Impact of changing the radii by lift control devices

The incorporated convex or concave surfaces of the aircraft wings are rotating the air masses creating centripetal forces towards the center of each convex or concave, thereby producing the useful lift forces. In many convex and concave surfaces, radius is a permanent feature, but flaps, slats and other lift control devices adjusts the total radius and the surface area significantly which results in increasing or decreasing the generated lift as depicted in Figure 4:9 and 4:10. In conventional explanations, the increased lift by control devices are attributed to the increased camber and the area of the wing. But, in accordance to the Don's Theory, increased lift is attributed primarily to the increment of the rotated air mass and decrease of the rotational radius.

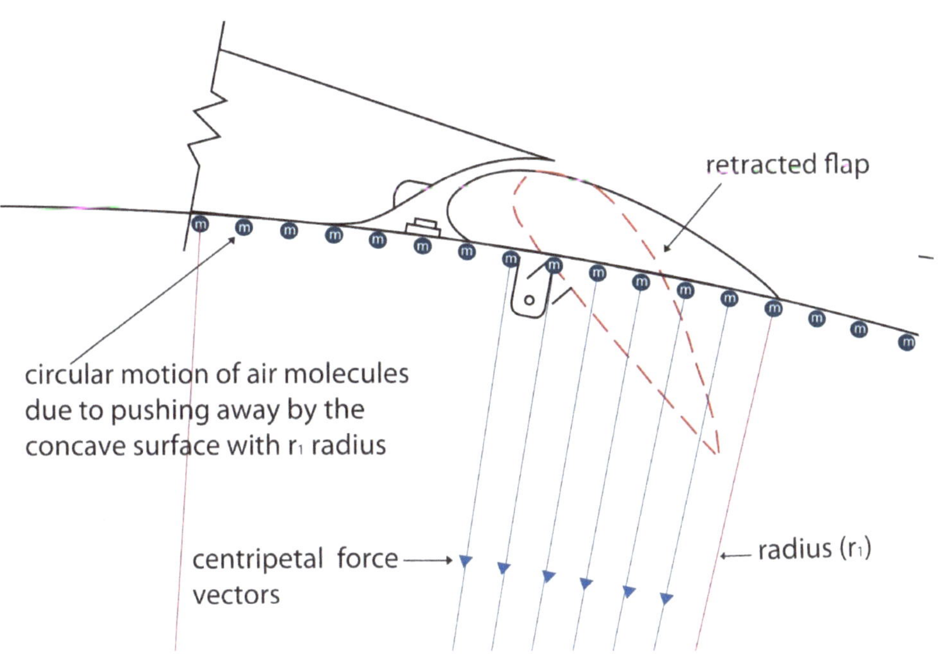

Figure 4:9 – retracted flap position

The deflected flap in Figure 4:10 transfer the circular motion of the air molecules that were previously circulating over a concave surface area with radius 'r_1' into circular motion with radius 'r_2'. Also, 'r_1' is approximately nine (9) times 'r_2' and also the mass of the deflected airstream is higher with the extended flaps.

However, the wing assembly depicted in Figure 4:10 fails to maintain the boundary layer with maximum efficiency due to its shape. The area coloured in blue could generate turbulence which reduces the efficiency of the molecular rotation. If the 'r_1' boundary layer could be transferred to the 'r_2' circular rotation without any loss or turbulence in boundary layer, the lift has to increase approximately by nine fold. This is the reason why the extension of flaps augment the lift considerably that the velocity of the aircraft could be reduced significantly during landing and takeoff.

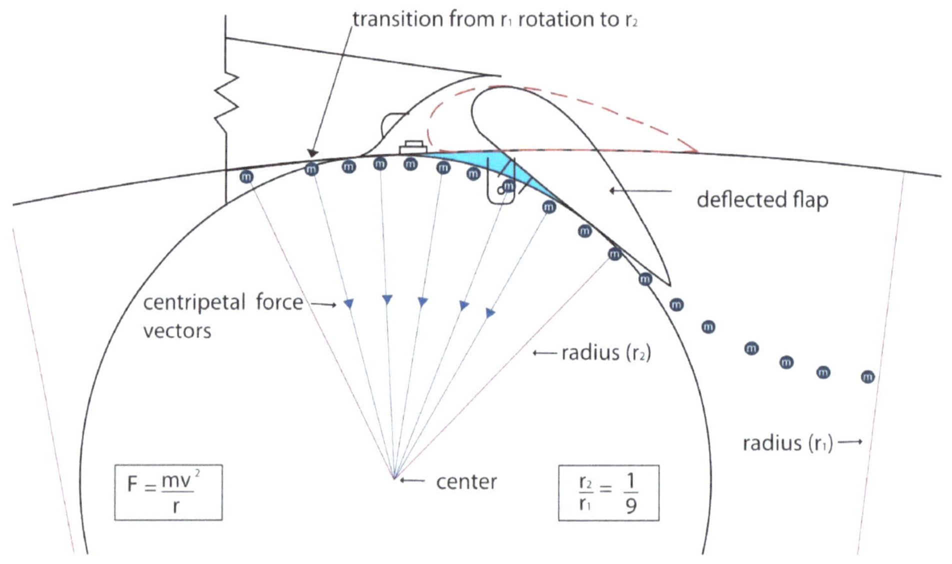

Figure 4:10 – deflected flap position

Orientation of convex or concave surfaces

The orientation of the centripetal force groups generated about an airfoil can either be upward (positive) or downward (negative). The sum total of the centripetal force groups (vector value) determines the total lift generated by a particular aerodynamic profile. In the aerofoil depicted in Figure 4:11, the segment '*AB*' and '*BC*' generate a positive lift at *10ºAoA*. The '*AD*' generates a negative lift while the segment '*DC*' generates a positive lift due to the '*coanda effect*'. Therefore, this aerofoil can be classified as a high lift profile.

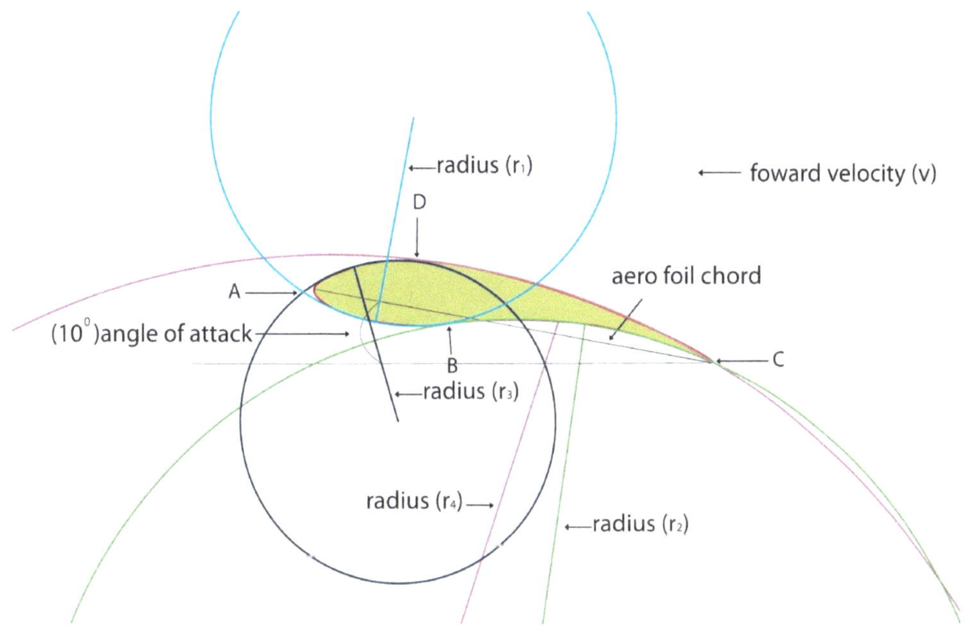

Figure 4:11 – high lift profile

Mass of the rotating air volume

Mass of the rotating air volume varies with in a limited range along with the *AoA*, but has a significant impact on the total lift generated. Some factors which influence the rotating air mass are

- Altitude

- Temperature

- Amount of water vapor or humidity

Angle of Attack (*AoA*)

The impact of the *AoA* to the generated lift are mainly two fold The increased *AoA* from 0^0 to 15^0 are influencing the lift in following manner.

- Increases the rotated air volume with increments of *AoA*, thereby increasing the lift.

- Negative circular surface orientation reduces by the increased direct exposure to the upcoming air stream, thereby increasing the lift with increments of AoA

Variation of the rotated air volume with AoA

The streamlined airflow which is fleeing backwards on both sides of the moving airfoil is varying its thickness which is directly proportional to the *AoA*. In a simpler explanation, as the *AoA* increases, the height (thickness) of the air stream deflected towards the rotation increases. The Figures 4:12, 4:13 and 4:14 indicates the increasing *AoA* from *0°, 5°* to *10°* are increasing the height (thickness) of the rotated air volume on both sides. The value of the thickness (height) is approximately equivalent to '*Chord x Sine (AoA)*' on the lower surface while the thickness of the rotated air volume is higher over the upper surface. The increased thickness of the rotated air stream results in increased volume and thereby, increased air mass. Hence, it can be concluded that the generated lift is directly proportional to the *AoA*. However, the increased air mass that requires to be rotated with the increasing *AoA* generates more drag which has to be compensated with increased engine power (thrust).

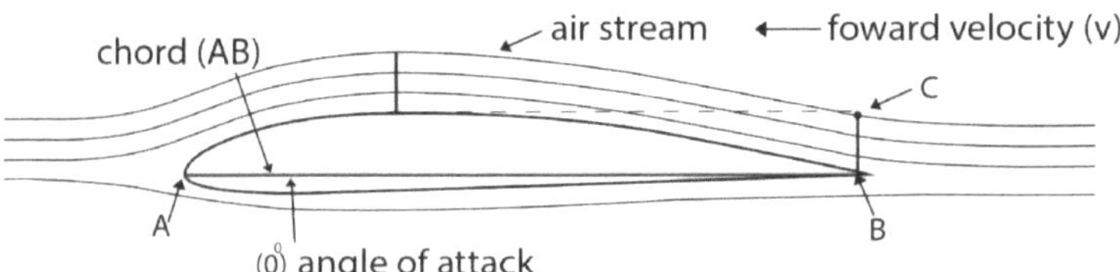

Figure 4:12 – Clark Y aerofoil at 0°AoA

The Figure 4:12 depicts a '*Clark Y*' aeorfoil set at the *AoA of 0°*. The thickness of the rotated airstream below the lower surface remains thin while the thickness of the rotated airstream is higher over the upper surface. The thickness of the airstream over the upper surface is approximately equivalent to highest camber point of the aerofoil from the trailing edge datum ('*BC*'). The upper airstream is pulled downwards and kept rotated after the highest camber point due to the '*coanda effect*'. The higher thickness of the airstream over the upper surface signifies the generation of higher amount of lift from the upper surface after passing the point with highest camber.

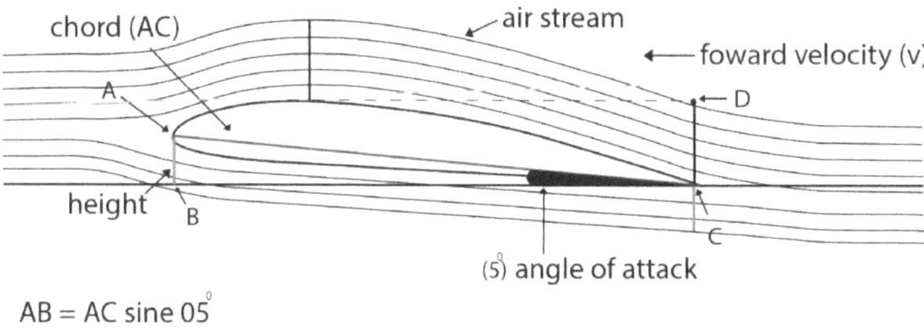

Figure 4:13 - Clark Y airfoil at 5°AoA

The Figure 4:13 depicts a '*Clark Y*' aerofoil set at the *AoA* of 5^0. The thickness of the rotated air volume below the lower surface is approximately '*chord x sine 5^0*' while the thickness over the upper airstream rotated is more than the thickness of the lower airstream. The thickness of the upper airstream is a function of the vertical depth that air molecules at the highest camber point speeds down to the datum level of the trailing edge rotating along the curved path.

Figure 4:14 depicts a '*Clark Y*' airfoil set at an *AoA* of *10⁰*. The thicknesses of the airstreams that are rotated by the convex/concave shapes of both lower and upper surfaces are increased compared to *5⁰* of *AoA*, thereby increasing the mass of the airstream and the resultant lift and the drag.

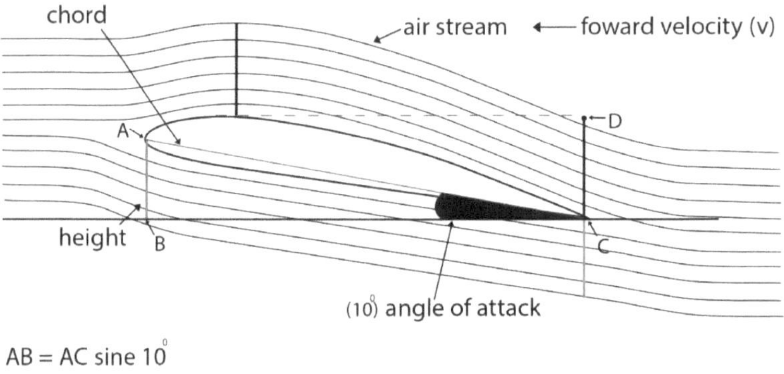

Figure 4:14 - Clark Y airfoil at 10°AoA

Boundary layer separation or stall (loss of coanda effect)

However, all literature related to experimental testing of the lift generation proportionately to increments in *AoA* indicates that after *15⁰* of *AoA* , the top layer of streamliners start cavitation from the trailing edge advancing on an aft-forward direction leading to a progressive separation of the laminar flow. The separation of the laminar flow closer to the trailing edge which is on a circular motion induced due to the *coanda effect* leads to the loss of significant lift or the phenomenon is known as the '*stall*'. The obvious reason for separation of the boundary layer and loss of '*coanda effect*' is the gradual increase of the thickness and the depth that the airstream has to be pulled down in order to maintain the laminar flow. The increase of the depth and thickness of the airstream over powers the suction force that pulls the airstream into circular motion as depicted in Figure 4:15. Once the laminar flow is disturbed, the vacuum gets filled up by the air molecules coming from other directions creating turbulence and thereby, stalling that segment of the wing.

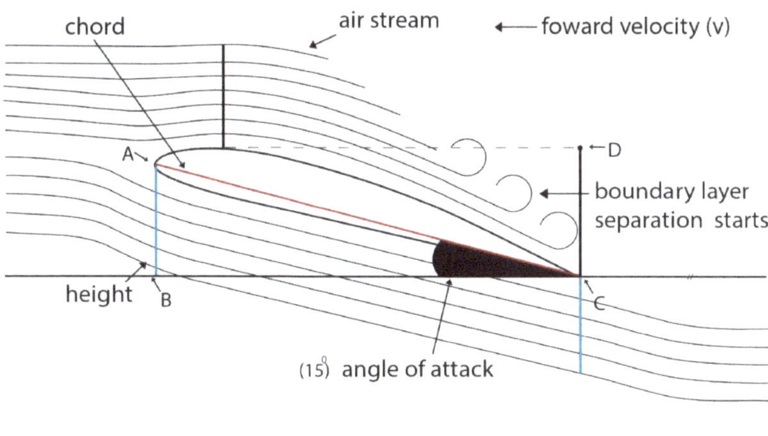

AB = AC sine 15°

Figure 4:15 – boundary layer cavitation after 15° of AoA

Change in surface area orientation relative to the AoA

In all aerofoils, the orientation of the certain convex or concave segments may vary with the *AoA* towards positivity or negativity, thereby making changes in the total positive lift generated. Figures 4:16 depicts the *'Clarke Y'* aerofoil set at $0°$, $5°$, $10°$ and *$15°AoA$*. It is observed that the convex segment *AD* which creates a negative lift at *$0°AoA$* is getting smaller as the *AoA* increases, thereby increasing the total lift. Also the *'BC'* segment that generates negative lift at $0°$, starts generating a positive lift as its orientation changes from suction to direct rotator of the airstream.

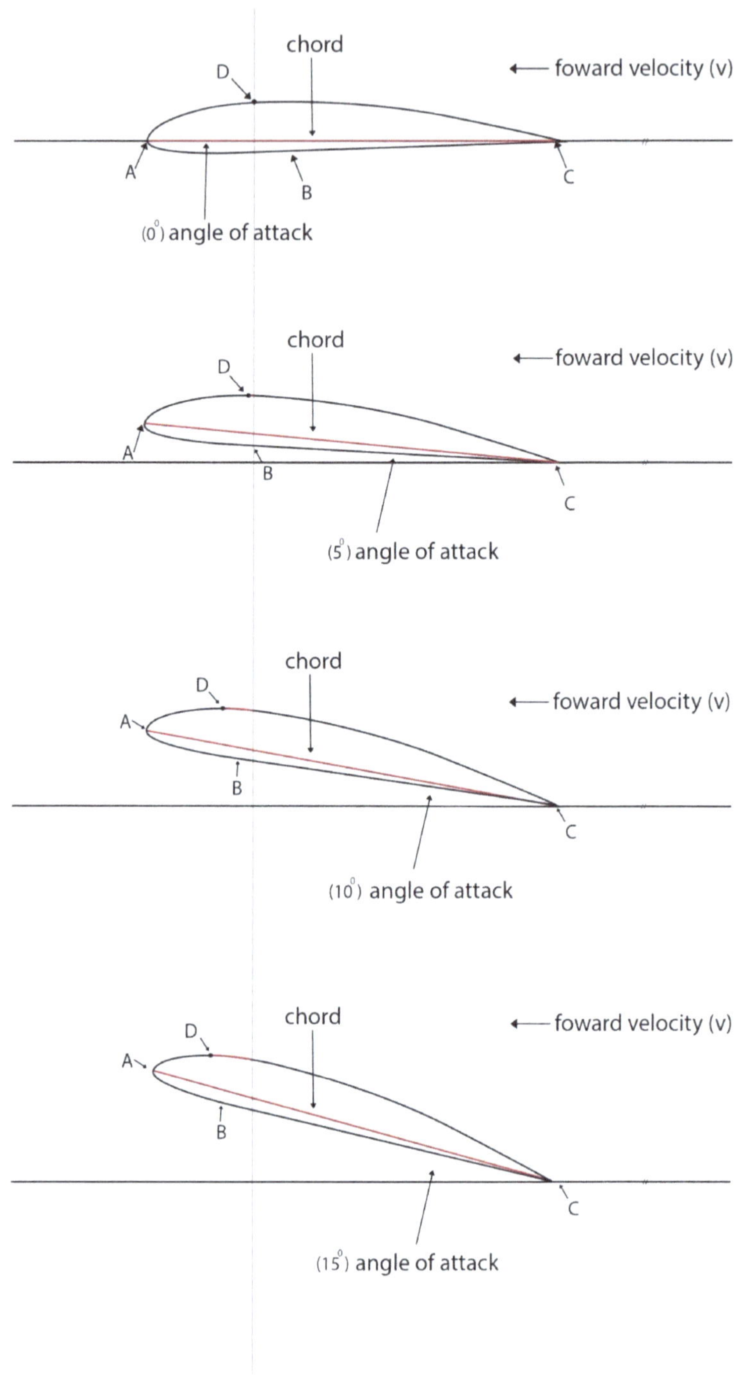

Figure 4:16 – shifts in orientation with AoA

'The vertical vector sum of centripetal forces induced due to the circular motion of air molecules in non-turbulent streamlines around convex and concave surfaces of an aerofoil is equal to the useful aerodynamic lift generated'.

A new theory to the world, termed as *'Don's Theory on Aerodynamic Lift'*. Different curves in different wings significantly changes the amount of upward lift force generated and that force can be accurately measured using the equations of circular motion without resorting to the experimental values of *'C_L'* as hitherto done. Similarly, the induced *'drag'* is equal to the horizontal vector sum of the centripetal forces generated

Therefore, *Don's Theory* will largely influence the way aircraft designers think and how pilots perceive the magic of flight in future. This theory will lead to new wing designs that can increase the efficiency and minimize the waste of useful energy in many folds with resultant optimization of the overall performance.

Difference between the Don's Theory and other lift theories

All aerodynamic lift theories from *'Bernoulli'* to *'Euler'* are based on the assumption that useful lift is generated by the differential pressure envelope over the top and bottom surfaces of the aircraft wings. It is believed that differential pressure envelopes are created due to the variations of the dynamic pressure which is directly proportional to the velocity of the air vehicle. Every effort has been taken to substantiate this perception using the *'Bernoulli Principle'*. Also, pressure is not a vector and therefore, it is difficult to orientate the generated lift force due to the application of differential pressure on wing surfaces. *'Don's Theory'* is completely disconnected from the idea of differential pressure and variable dynamic pressure. *'Don's Theory'* explains the lift as the vertical vector sum of all centripetal forces generated by the circular motion of air masses above and beneath an aircraft wing. Hence, the rotated air mass, velocity of rotation, orientation and the radii of the convex or concave surfaces are the four main variables of the generated aerodynamic lift according to the '*Don's Theory*'. Rotated air mass is directly proportional to the *AoA* of the wing, air density and wing area.

Quantification of lift force by the application of Don's Theory

An aircraft wing may consist of multiple circular surfaces that could rotate the air masses which are fleeing past them. Therefore, the following sequencing can improve the accuracy of the quantification.

a.   Segregation of the convex and concave surfaces about an aerofoil and establish the radii of each curvature.

b. Calculation of the air mass which is being induced into circular motion by each airfoil segment (*volume x density = mass*) which is a function of the AoA.

c. Calculate the centripetal force generated by each circular segment of the airfoil.

d. Calculation of the vector value of the upward centripetal forces considering the orientation of each force segment (force vectors).

The generated force could be positive or negative depending on its orientation as depicted in Figure 4:16. Therefore, the useful aerodynamic force is equivalent to the vertical vector sum of the total centripetal forces generated.

Chapter Five

APPLICATION OF DON'S THEORY TO DIFFERENT AEROFOILS

Clark Y

This profile had been designed in 1922 by *Virginius E. Clark* and was widely used in general purpose aircraft designs, and much studied in aerodynamics over the years. The flat bottom simplifies the angle measurements on propellers, and makes it easy for construction of wings on flat surfaces. For many applications, the *'Clark Y'* has been a satisfying airfoil section and it has reasonable overall performance in respect of its *lift-to-drag* ratio. It's gentle and relatively benign stall characteristics can also be considered as benefits, but is rarely used in modern designs. According to the *'Don's Theory'* flatter the surface, lesser the lift due to the high values of the radii.

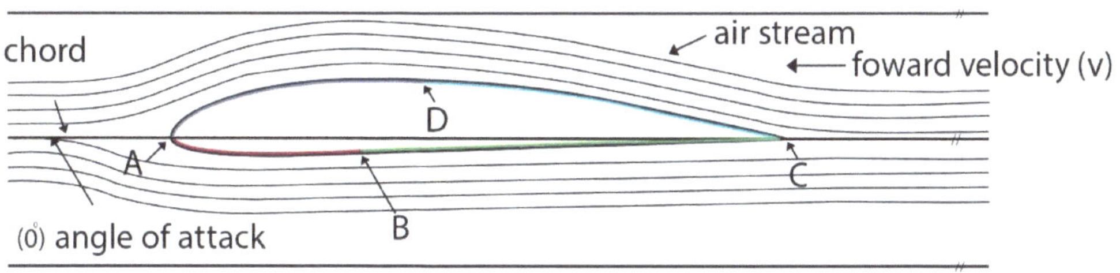

Figure 5 : 1 - Clark Y Profile at 0°AoA

Figure 5:1, indicates a *'Clark Y'* aerofoil set at an AoA of 0° and moving forward with a velocity *'v'*. Along the lower and upper contours of the aerofoil, different convex and concave surfaces can be identified. They have different lengths, orientations and radii. The transect points of each circular segment on the profile can be marked as 'AB', 'BC', 'AD' and 'DC'. Figures 5:2 through 5:5 discuss about the quantification of the generated lift from each segment and their orientation in accordance with the *'Don's Theory'*.

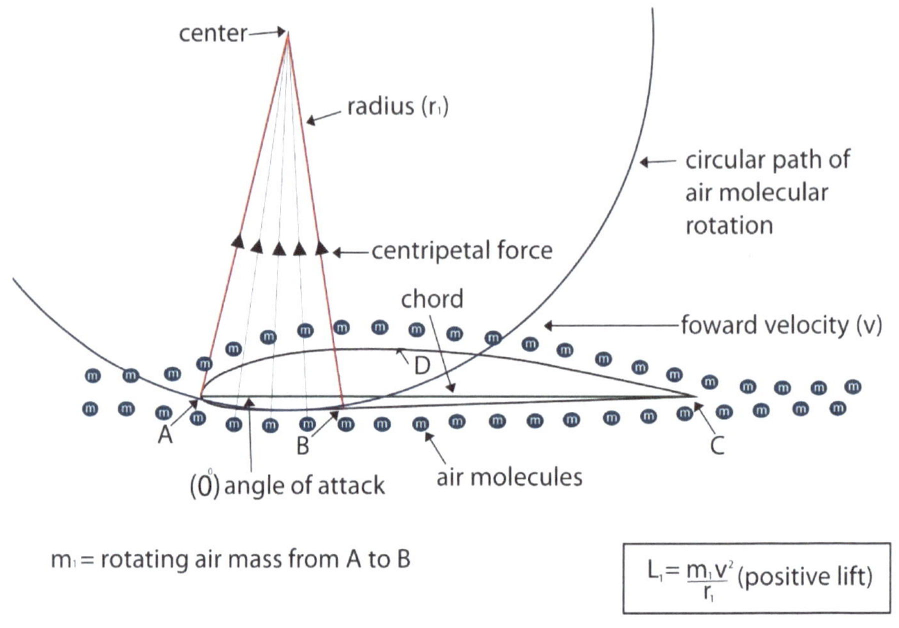

Figure 5:2 - arc AB on the circle with radius r₁

The segment 'AB' on the lower surface of the aerofoil is clearly an arc of the circle with radius *(r₁)* as depicted in Figure 5:2. At AoA of 0⁰, the convex 'AB' is pushing the airstream on a downward direction and sending it on a circular motion from point 'A' to 'B'. Therefore, the air molecules that are rotating about the circumferential distance (arc) 'AB' at a velocity of *'v'* and having a mass of *'m₁'* generates a resultant centripetal force of *'L₁'*. 'AB' arc of the airfoil is downward looking convex profile and therefore, it holds the air stream (the stream of air molecules) on its circular path by exerting an *'equal and opposite reactive force'* against the centripetal forces vectored towards the center of the circle. Therefore, the vector value of *'centripetal forces'* exerted by the *'m₁'* air mass rotating over segment 'AB' of the aerofoil is directly equivalent to the generated *'positive lift'*.

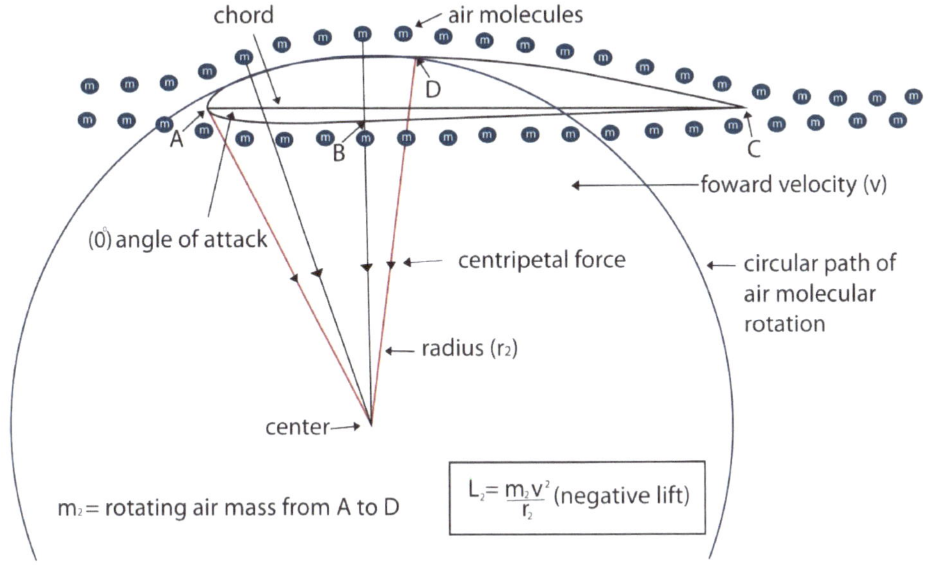

Figure 5:3 - arc 'AD' on the circle with radius 'r_2'

The segment 'AD' of the aerofoil depicted in Figure 5:3 is a convex profile oriented towards a direction so that it holds the air stream on a circular orbit with the radius 'r_2' which is approximately opposite to the segment the 'AB'. In another perspective, 'AD' segment is pushing the airstream on an upward direction from its original path exerting an *'equal and opposite reactive force'* against the centripetal forces directed towards the center of the circle with radius 'r_2'. The air molecular mass *(m₂)* that is fleeing over the circular path 'AD' generates centripetal forces that are applying force on the segment 'AD' directly on a downward direction creating a *'negative lift'* (L₂).

A significant disadvantage of the *Clark Y* profile is the lower radius of the upper segment 'AD' over the bottom segment 'AB'. This is a result of the flat bottom configuration due to which the *'negative lift'* generated by the upper segment 'AD' is higher than the 'positive lift' generated by the bottom segment 'AB'. Therefore, the resultant vector value is 'negative lift' ($L_1 < L_2$) for the frontal area of the profile at AoA of 0^0.

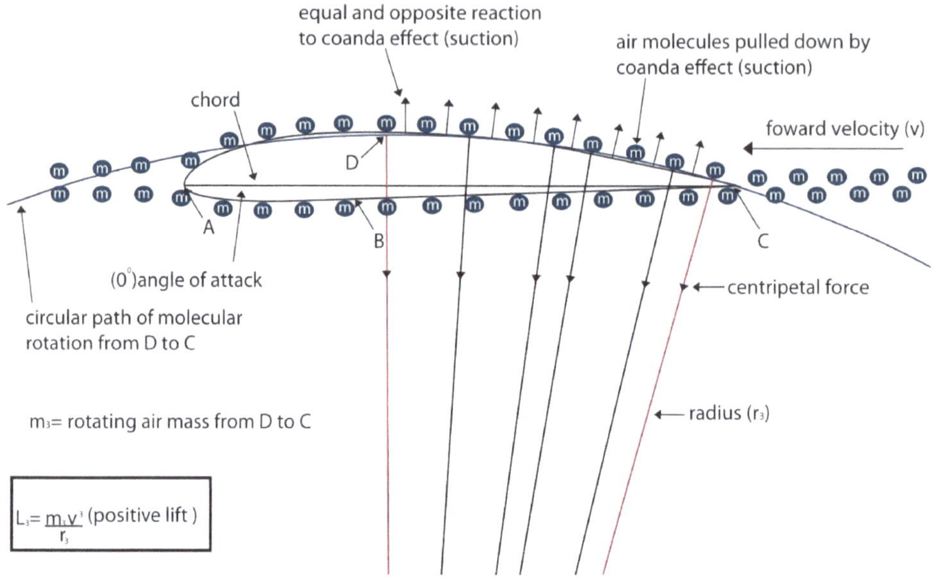

Figure 5:4 - arc DC on the circle with radius r_3

The stream of air molecules fleeing over the upper surface of the *'Clark Y'* aerofoil depicted in Figure 5:4 takes a downward turn after the point 'D' along the circumferential distance 'DC' due to the *'Coanda Effect'*. This phenomenon takes place by the *'pulling force'* exerted on the *'m₃'* air mass by the segment 'DC' due to suction. As the air molecules are pulled down, the 'DC' surface is drawn up with an equal and opposite reactionary force creating a *'positive lift'*.

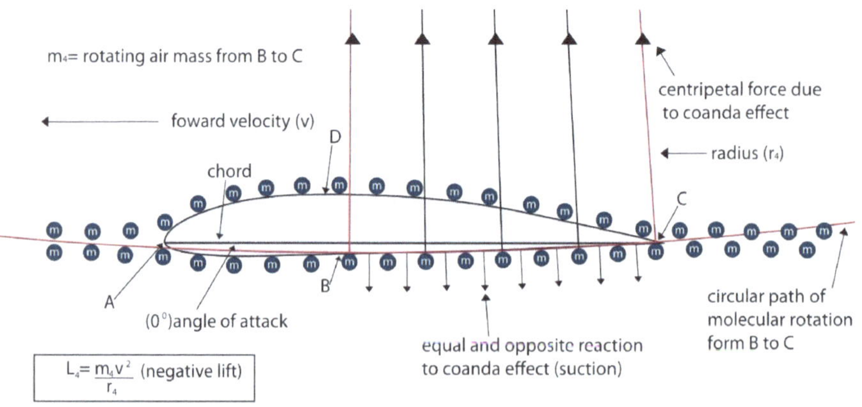

Figure 5:5- arc BC on the circle with radius r_4

Figure 5:5 indicates the circular motion of the air mass *'m₄'* along the 'BC' segment of the aerofoil. At 0°AoA, the seemingly flat area after the point 'B' is slightly inclined upwards and therefore, the airstream is sucked up due to the *coanda effect*. The flat surface area 'BC' could be considered an arc of a circle that has a very high radius *(r₄)*. Therefore, the 'BC' segment of the *'Clarke Y'* aerofoil generates a negative lift or no lift due to the high value of radius *(r₄)* and the slight inward inclination towards the trailing edge at AoA of 0°.

Therefore, the total *Lift (L)* = *positive lift (L₁ + L₃)* − *negative lift (L₂+ L₄)*

$$L = \left(\frac{m_1v^2}{r_1} + \frac{m_3v^2}{r_3}\right) - \left(\frac{m_2v^2}{r_2} + \frac{m_4v^2}{r_4}\right)$$

'Clark Y' aerofoil inverse

Since *'Clark Y'* is a popular aerofoil profile, it is interesting to see what happens if it flies inverse. AoA is set at 0° and now not only the 'AB' segment is pushing the air molecules into circular motion, it also pushes them slightly upwards. The vector sum of the centripetal forces directed towards the center of the circle with radius *'r₁'* is equal to the 'negative *lift*' on 'AB' convex as depicted in Figure 5:6.

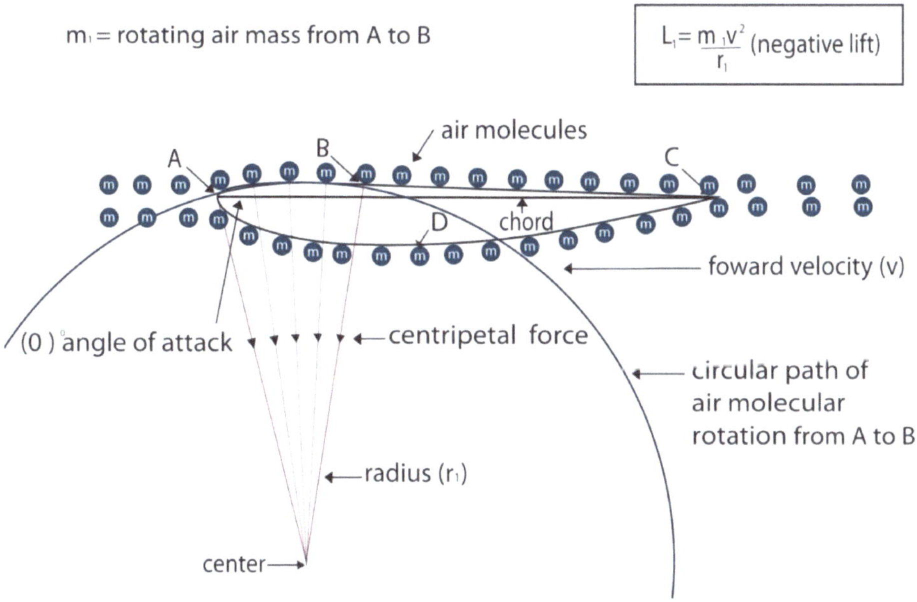

Figure 5:6 – negative lift induced by the 'AB' segment

Figure 5:7 indicates that the polarity of the centripetal forces generated by the rotating air mass along the 'AD' convex is projected upwards. Hence, the exerted centripetal forces on the air molecules that are taking the circular path from 'A' to 'D' can be reckoned as *positive lift* acted upon the 'AD' convex.

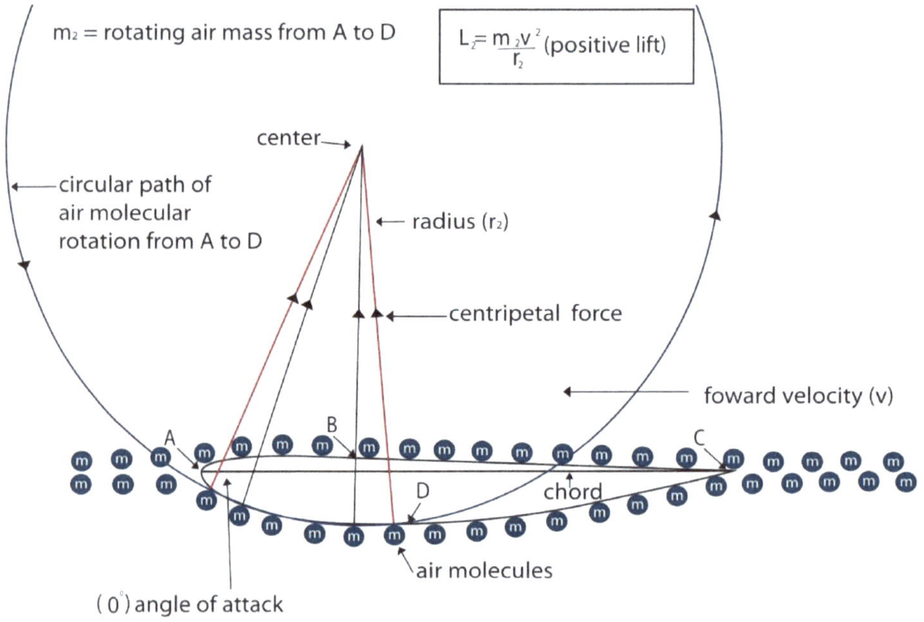

.Figure 5:7- positive lift induced by the 'AD" segment.

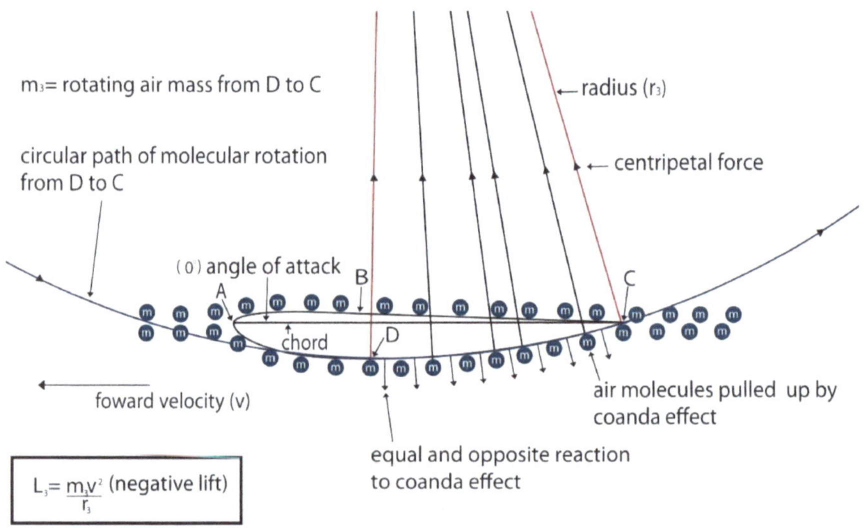

Figure 5:8- negative lift induced by the 'DC' segment.

The air molecules that are travelling along the circular path 'DC' remain as a boundary layer due to the *coanda effect* as depicted in Figure 5:8. Molecules that are pulled up by the *coanda effect* or the suction created due to the vacuum are in equilibrium with the equal and opposite downward suction force exerted on the 'DC' segment which has to be reckoned as *'negative lift'*.

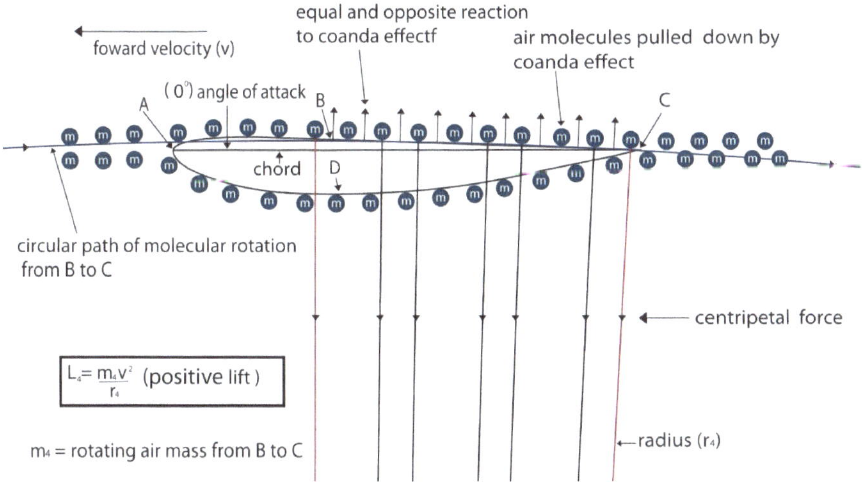

Figure 5:9–positive lift due to coanda effect

Air molecules that are rotating along the segment 'BC' (arc) which has a radius 'r_4' remains attached to the 'BC' surface due to the *coanda effect*. The *coanda effect* generates the centripetal forces which keeps the air molecules on a circular motion over the convex distance 'BC' with a radius 'r_4'. However, the value of 'r_4' is much greater in comparison to 'r_1','r_2' and 'r_3' (due to the flat surface 'BC') which makes the value of 'L_4' significantly smaller or negligible. *Coanda effect* or the suction force exerted on the 'BC' segment is equal and opposite to the centripetal forces exerted on the air molecules. The orientation of the suction force could be reckoned as *'positive lift'*, though negligibly small due to the high value of 'r_4' as depicted in Figure 5:9.

Therefore, the total Lift *(L)* = *positive lift (L₂ + L₄) – negative lift (L₁ + L₃)*

$$L = \left(\frac{m_2 v^2}{r_2} + \frac{m_4 v^2}{r_4}\right) - \left(\frac{m_1 v^2}{r_1} + \frac{m_3 v^2}{r_3}\right)$$

The ratios of the r_1, r_2, r_3 and r_4 extracted from the scaled drawings,

r_1(negative) : r_2(positive) : r_3(negative) : r_4 (positive) =1.00 : 1.15 : 2.70 : 21.09

Above ratios indicate that the positive lift generated from the convex surfaces with 'r_2' and 'r_4' is outweighed by the negative lift generated from the convex surface segments with 'r_1' and 'r_3' due to the comparatively low values. Hence it is evident that the '*Clark Y aerofoil inverse*' generates a *'negative lift'* at AoA of 0°.

NACA 0018 (Symmetrical AeroFoil)

The Figure 5:10 depict the impact of the centripetal forces on the NACA 0018 Symmetrical airfoil at AoA of 0^0. Owing to the symmetric shape of the airfoil $r_1 = r_2$ and $r_3 = r_4$. Also the orientation of 'AB' is exactly opposite to the 'AD' and 'BC' is the opposite of 'DC'. Therefore, the centripetal forces generated about a symmetrical airfoil at AoA of 0^0 cancel each other generating a zero lift.

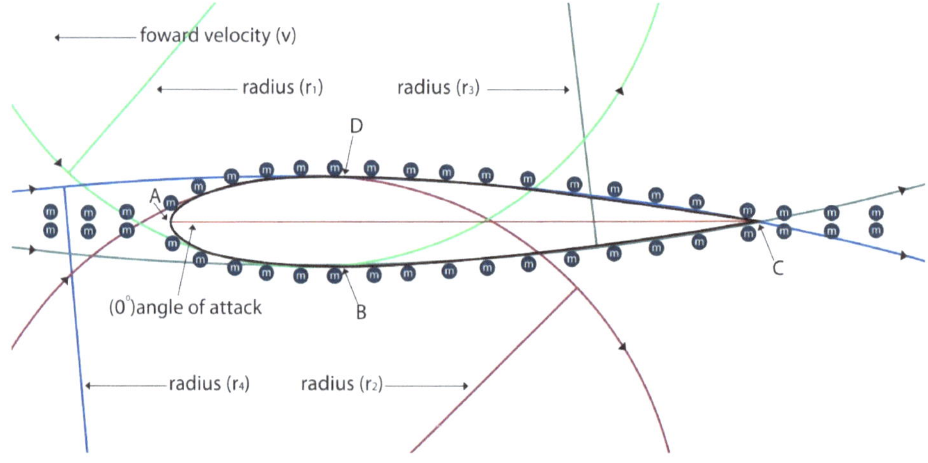

Figure 5:10 – NACA 0018 at 0^oAoA *(r_1= r_2 and r_3 = r_4)*

Lower surface of the NACA 0018 at 10°AoA

As the AoA increases, the airstream starts directly confronting the 'AB' and 'BC' segments of the aerofoil. Therefore, at the lower surface, the airstream is being forced into circular motion around the 'AB' and 'BC' convex as shown in Figure 5:11. Also, the 'BC' convex which induced a negative lift owing to the *coanda effect* at AoA of 0° is now producing a positive lift due to the direct confrontation of the airflow pushing it into circular motion around the convex 'BC' as indicated in Figure 5:12.

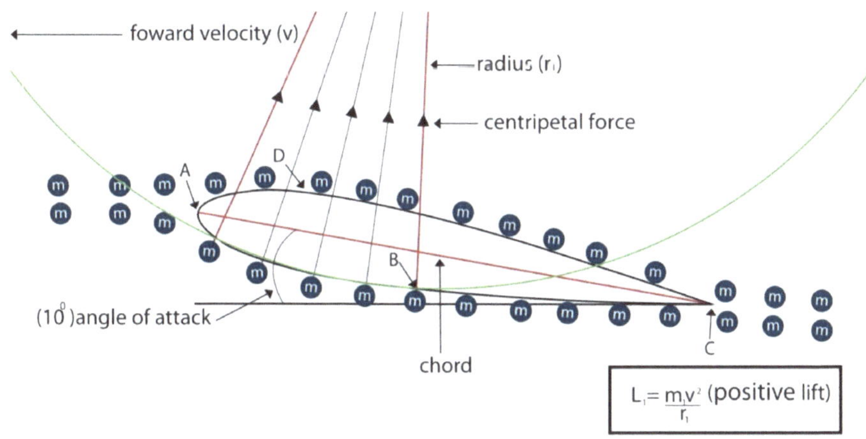

Figure 5: 11 – AB segment of NACA 0018 at 10°AoA

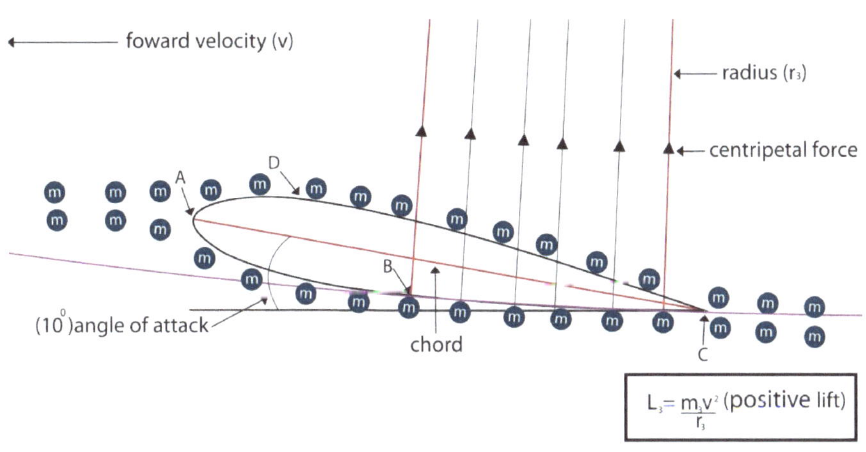

Figure 5: 12- BC segment of NACA 0018 at 10°AoA

Upper surface of the NACA 0018 at 10⁰AoA

As the AoA increases, the length of the 'AD' convex which pushes the airstream into circular motion is reducing gradually by the advancement of the point 'D' forward. 'D' is the last point that directly deflects the upcoming airstream in to circular motion, before it is pulled down along the 'DC' segment as a boundary layer (for AoA below 15⁰). The advancement of point 'D' forward is reducing the negative lift induced from 'AD' convex as depicted in Figure 5: 13.

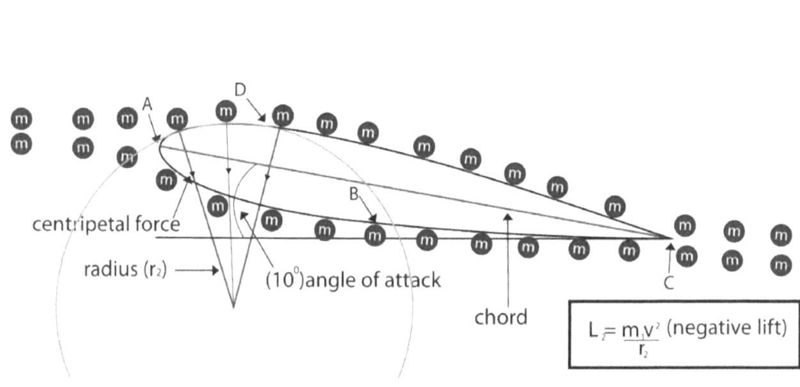

Figure 5:13 – only negative lift segment of NACA 0018 at 10 ⁰

As the upper airstream passes the point 'D', the *coanda effect* pulls the airstream down and it remains as a boundary layer while rotating around the convex 'DC' with radius *'r₄'* as depicted in Figure 5:14. Radius *'r₄'* is comparatively much higher than *'r₂'*, but is a constant that does not change with the changing AoA. The mass of the airstream that pulls down by the *coanda effect* is significantly increased with the increasing AoA. The equal and opposite reaction to the downward pull of the airstream is the upward suction of the 'DC' convex which is a positive lift.

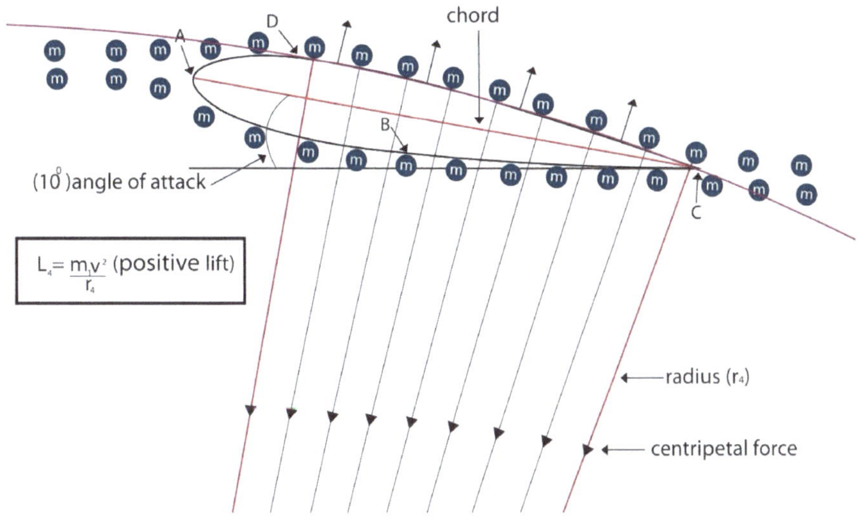

Figure 5:14 – positive lift due to coanda effect in NACA 0018 at 10 ^{o}

Therefore, the total Lift *(L)*   = *positive lift (L₁+ L₃+L₄) – negative lift (L₂)*

$$ L \;\; = \left(\frac{m_1 v^2}{r_1} + \frac{m_3 v^2}{r_3} + \frac{m_4 v^2}{r_4} \right) - \left(\frac{m_2 v^2}{r_2} \right) $$

'Don's Theory on Aerodynamic Lift' clearly explains why a symmetrical aerofoil does not produce any lift at AoA of 0^0 and alternately, how it produces a significant lift at AoA of 10^0. In a like manner *'Don's Theory'* can be applied to any aerodynamic profile to simply analyze their effectiveness and behavior at different AoAs in order to customize them. Performance of profiles at different orientations and speeds could be determined with much ease and understanding the actual dynamics behind the application which is more meaningful than mere CFD analysis. Application of *'Don's Theory'* can give a new meaning to the results of CFD analysis and interpret the phenomenon with a credible theory.

BIBLIOGRAPHY

1.   John D. Anderson Jr.(1997), *A History of Aerodynamics and its impact on Flying Machines*, Cambridge University Press, Britain

2.   Theodore Von Karman, (2004), *Aerodynamics: Selected Topics in the light of their Historical Development*, Courier Corporation, USA

3.   EL Houghton, PW Carpenter, Steven Collicott and Daniel T Valentine, (2013) *Aerodynamics for Engineering Students*, Butterworth-Heinemanne, USA

4.   Lazar Dragos,(2003),*Mathematical Methods in Aerodynamics*, Kluwer Academic Publishers, Boston- London.

5.   John D. Anderson Jr,(1997), *Fundamentals of Aerodynamics*, Tata McGraw Hill Education Private Limited, New Delhi.

6.   Herman T.Schlichting and Erich Truckenbrodt (1979), *Aerodynamics of the Airplane*, Tata McGraw Hill Education Private Limited, New Delhi.

7.   Holt Ashley and Marten Landahl (1965), *Aerodynamics of Wings and Bodies*, Dover Publications, INC, New York.

8.   SP Langley (2009), *Experiments in Aerodynamics*, Bilio Bazaar, New Delhi.